万达商业规划

持有类物业 上册 VOL.1

WANDA COMMERCIAL PLANNING 2014 PROPERTIES FOR HOLDING

2014

万达商业规划研究院 主编

中国建筑工业出版社

EDITORIAL BOARD MEMBERS
编委会成员

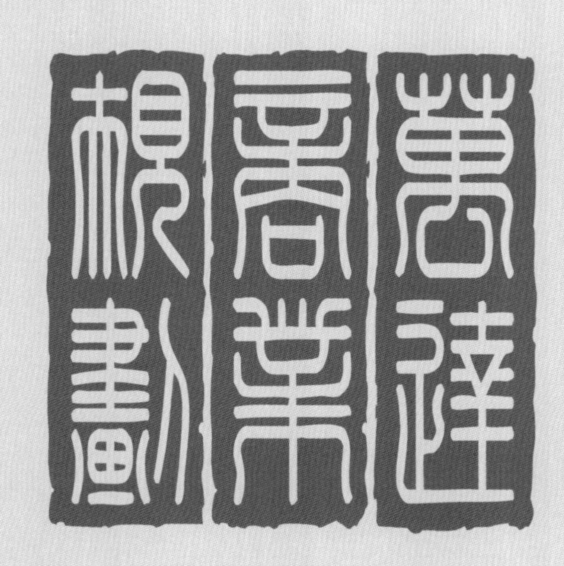
萬達
商業
規劃

万达广场
vanguard

达广场
NDA PLAZA
非机动车
停放点

CONTENTS
目录

C WANDA HOTELS 万达酒店 140

D WANDA CINEMA 万达影院 196

E DESIGN AND CONTROL 设计与管控 208

万达广场

FOREWORD
序言
WANDA
COMMERCIAL
PLANNING
2014

CHAIRMAN WANG JIANLIN TALKING ABOUT REAL ESTATE TRANSITION
王健林董事长谈房地产转型

现在，从事房地产行业和准备还要从事这个行业的人都要好好想一想，绝对不能再走传统路子，要抓住最后十年的机会转型升级。我们研究了世界许多发达国家，也包括我国台湾和香港地区，发现世界上任何一个国家和地区的房地产发展都很难超过半个世纪，一般就是三四十年的繁荣期，城市化进程达到70%到80%，这个行业的规模就开始萎缩。有人说，中国房地产黄金十年过去了，还有白银十年，我不说是黄金还是白银，我认为留给大家的转型时间也就还有十年罢了。这个行业的从业者，可能都不希望自己企业只活十几年，都希望自己企业成为百年企业，那就要用好这十年时间转型。如果中国城镇化达到70%，可能机会就没有了。中国不像美国、不像欧洲，城镇化率可以达到80%以上，中国还有大量山地，不可能把山上的地都拿出来搞建设。所以，希望房地产企业加快转型，万达自己也在加快转型。

——摘自王健林董事长2014年12月13日在第十三届中国企业领袖年会上所作《房地产的新常态》主题演讲

万达集团董事长
王健林

Now, it's the high time for real estate practitioners and those who want to enter real estate industry to think hard about how to grasp the next and last ten years as an opportunity to conduct transition and upgrading with the traditional approaches being futile and out of the question. After studying the former experiences of many developed countries in the world, including Chinese Taiwan and Chinese Hong Kong, we find that in any countries and regions of this world, it's hard for real estate development to last for more than half a century. Normally, the booming period shall last for only 30 to 40 years, and when the urbanization process reaches 70% to 80%, the industry scale will begin to shrink. Some people say that after the "golden decade" of Chinese real estate, there is still the "silver decade". I am not talking about gold or silver here, what I want to say is that, I think there are only at most ten years left for us to make transition. For practitioners in this industry, I believe that it is unacceptable for all that your company shall only survive for only a dozen years. Therefore, if you want your company to become a "Centennial Enterprise", the next ten years shall be taken good use of to conduct transition. Once the urbanization rate in China reaches 70%, the opportunity shall be gone. It's unlikely for China to reach an urbanization rate of more than 80% like the US and Europe. China is rich in mountain land, and it's impossible to conduct construction on mountain land. Therefore, I expect real estate enterprises to quicken their transition speed, and Wanda is also making transition at a fast speed.

—Excerpted from Chairman Wang Jianlin's speech titled *"The New Normal of Real Estate"* delivered in the 13th China Entrepreneur Summit on 13th December, 2014

Chairman of Wanda Group
Wang Jianlin

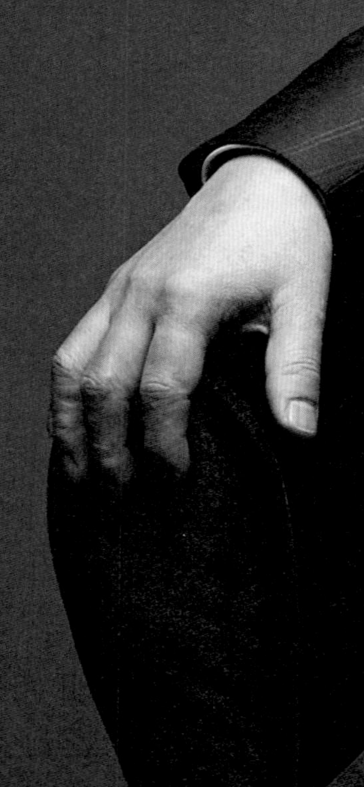

WANDA COMMERCIAL PLANNING 2014
万达商业规划 2014

万达商业地产副总裁　赖建燕

“2014年12月23日上午9点30分，伴随着香港联交所的一阵钟声，万达商业地产成功在港交所主板挂牌上市。”

"With a bell tolled in Hong Kong Stock Exchange, Wanda Commercial Properties Co. Ltd. completed a successful listing on the main board of the Hong Kong Stock Exchange at 9:30 a.m. on 23rd December 2014".

或许，在此之前，有很多人可能尚未关注万达，但这一刻足以吸引全球的目光。万达以“全球第二大不动产经营商”、开业广场109座、开业酒店71座、拥有院线屏幕数1616块等一系列令人骄傲的数字，成为“港交所”2014年最大规模的IPO。正如王健林董事长所讲，“2014年是万达集团的标志年”。

Before that moment, many people might have not paid close attention to Wanda yet. However, this was definitely the moment that attracted attentions worldwide. Wanda, being "the second largest real estate operator in the world" and the proud owner of 109 operating plazas, 71 operating hotels and 1616 cinema screens in Wanda Cinema Chain, became the largest IPO of the year 2014 in Hong Kong Stock Exchange. As Chairman Wang Jianlin said, "2014 is a landmark year for Wanda".

2014年是万达集团的标志年

2014 IS A LANDMARK YEAR FOR WANDA

一、标志性项目

I. LANDMARK PROJECTS

2014年也是万达商业规划的标志年，这里不得不首先提到2014年开业的三类不同类型的标志性建筑。

2014 is also a landmark year of Wanda's Commercial Planning. Here, it is inevitable to firstly mention the opening of three different types of landmark buildings in 2014.

1.标志一 ——第100座万达广场开业

昆明西山万达广场地处昆明市泛亚金融产业中心园区，包括商业中心、城市步行街、超五星级酒店和超5A甲级写字楼等，占地面积7.0公顷，总建筑面积70万平方米。其中，超五星级万达文华酒店和300米高“昆明双塔”甲级写字楼建成后成为昆明金融产业中心园区乃至昆明全市的标志性建筑（图1）。

王健林董事长表示，100座万达广场的开业意义非凡，它不仅代表企业规模，也标志着万达广场成为具有世界影响的品牌。“万达广场就是城市中心”不仅是一个口号，也是真实能为百姓提供美好生活体验的地方。

2.标志二 ——第一座万达瑞华酒店开业

由万达商业规划院、万达文旅设计院、万达酒店院设计的武汉万达瑞华酒店位于武汉市武昌区东湖路，是万达首家七星级酒店，其奢华程度可与世界任何一家顶级酒店比肩。独具特色的“钻石”立面造型与“红灯笼”秀场相映生辉，让它一举摘取中国旅游业界2014年度“最佳新开业奢华酒店”的桂冠（图2）。

1. "LANDMARK ONE": THE OPENING OF THE 100TH WANDA PLAZA

Kunming Xishan Wanda Plaza, located in Kunming Pan-Asia Financial Industry Park, consisting of a business center, urban walking streets, a super-five-star hotel, premium 5A office buildings, etc., covers an area of 7.0 hectares, with an overall GFA of 700 thousand square meters. Among which, the super-five-star Wanda Vista and the 300-meter-

（图1）昆明西山万达广场

（图2）武汉万达瑞华酒店

3.标志三 ——首个城市中央文化区建成

从2009年开始规划、到2014年底武汉“汉秀”剧场和武汉电影科技乐园的开业，武汉中央文化区历经5年建设完成。它位于武汉市核心地段——武昌区东湖和沙湖之间——项目规划区域约1.8平方公里、总建筑面积340万平方米，是万达集团投资500亿元人民币倾力打造，以文化为核心兼具旅游、商业、商务及居住功能为一体的世界级文化旅游项目（图3）。

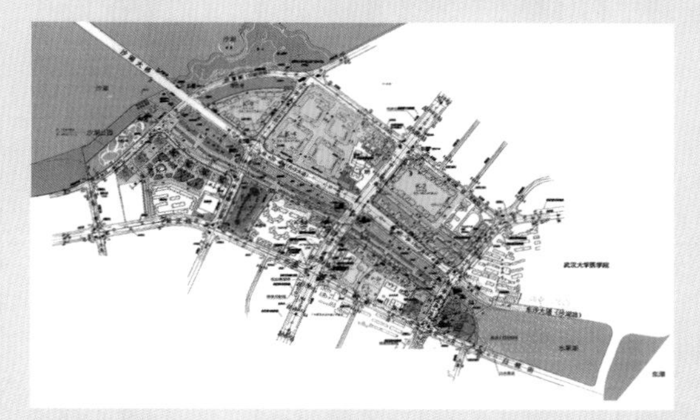

（图3）武汉中央文化区总图

二、模式转型

2014年作为万达集团标志年的又一重要标志是“模式转型”。这一转型主要体现在两个方面。

1.管理模式转型

（1）集团总部职能向项目公司下放

商业规划系统已完成创新管理模式培训并全面推行。未来规划系统将把更多精力投入在研发、创新等更有挑战、更有价值的工作上。

（2）全面实行“总包交钥匙”管控模式

“总包交钥匙”模式是一项革命性的创新之举，不仅有利于推进企业的反腐败工作，而且能实现多赢局面，使万达商业地产建设“生态链”更加健康。今后，所有新开工综合体项目将全面实行“总包交钥匙”管控模式。这是中国工程建设领域的一项革命性创新之举，也标志着万达商业地产的项目管控达到国际水准，并将对整个地产行业起到示范和引领作用。

high "Kunming Twin Towers" grade-A office buildings upon completion shall become landmark buildings in Kunming Financial Industry Park and even in the whole Kunming (as shown in Fig.1).

As Chairman Wang Jianlin says, the opening of the 100th Wanda Plaza has a significant meaning, which does not only reveal the enterprise scale, but also marks Wanda Plaza's becoming a brand of worldwide influence. Wanda Plaza, dubbed as the "Center of the City", is truly a place that can offer citizens an experience of good life rather than merely being a slogan.

2. "LANDMARK TWO": THE OPENING OF THE FIRST WANDA REIGN HOTEL

Designed jointly by Wanda Commercial Planning Institute, Wanda Cultural Tourism Planning & Research Institute and Wanda Hotel Design Institute, Wanda Reign Wuhan, located in Donghu Road, Wuchang District, Wuhan, is Wanda's first seven-star hotel, whose luxury can compete with any top hotel in the world. With its unique "diamond" façade modeling and the splendid "Red Lantern" show theatre, which enhance each other's beauty, Wanda Reign won the laurel of "Best Newly-opened Luxury Hotel 2014" awarded by China Travel & Meeting Industry Awards (as shown in Fig.2).

3. "LANDMARK THREE": THE COMPLETION OF THE FIRST CITY CENTRAL CULTURAL DISTRICT

From the initial planning in 2009 to the opening of Wuhan Han Show Theater and Wuhan Wanda Movie Park in 2014, it took 5 years to finally complete the construction of Wuhan Central Cultural District. the project is planned to cover an area of 1.8 square kilometers, with 3.4 million square meters of overall floorage. With an investment of 50 billion yuan by Wanda Group, it is a culture-centered world-class culture and tourism project with the integrated function of tourism, business, commerce and residence (as shown in Fig.3).

II. TRANSITION OF MODEL

Another important symbol for 2014 being a landmark year for Wanda Group is its "transition of model", which mainly represents in the following two aspects.

1. TRANSITION OF MANAGEMENT MODEL

(1) The delegation of power from Wanda headquarter to local project companies

The innovative management model training for the commercial planning system has been completed and the model has been fully implemented. In the future, the planning system shall put more energy into more challenging and valuable work, like research, development, innovation, etc.

(2) The full implementation of "Turnkey Contract" management and control model

The "Turnkey Contract" model is a revolutionary measure of innovation. Its implementation not only improves the anti-corruption work of Wanda, but also helps to achieve a multi-win situation, making the "ecological chain" of Wanda Commercial Properties construction much healthier. In the future, the "Turnkey Contract" management and control model will be fully implemented to all newly-opened complex projects. This is a revolutionary and innovative measure in the domain of China project construction, which indicates that the project management and control competence of Wanda Commercial Properties has reached international standard and shall set an example for and lead the whole industry.

2. TRANSITION OF PRODUCT MODEL

(1) Shopping centers

2.产品模式转型

（1）购物中心

集团为顺应“O2O”模式对传统商业模式的影响，对万达购物中心产品模式进行了相应调整，主要体现在：

- 百货：百货业态占比减少；
- KTV：大歌星娱乐业态占比减少；
- 儿童娱乐：增设儿童娱乐业态占比，打造宝贝王自创品牌。

（2）酒店

根据集团发展规划和市场情况，万达酒店产品模式进行了相应调整，主要体现在：

- 优化酒店餐饮结构，减少特色餐饮面积，中餐面积；
- 缩减城市酒店规模，减少客房钥匙数。通过产品调整，精简酒店规模，努力提升酒店品质，打造精品。

To comply with the influence brought to traditional business model by "O2O" model, corresponding adjustments have been made towards the product model of Wanda shopping centers, which mainly reflect in the following aspects:

- Department store: the percentage of department store business has been reduced.
- KTV: the percentage of "Super Star KTV" entertainment business has been reduced.
- Kids entertainment: kids entertainment business has been increased, with the establishment of proprietary brand "Kids place".

(2) Hotels

In accordance with the development plan of Wanda Group and market conditions, corresponding adjustments have been made towards the product model of Wanda hotels, which mainly reflect in the following aspects:

- Catering structure has been optimized, with a reduction of specialty catering area and Chinese restaurant area.
- Hotel scale has been downsized and number of room keys has been reduced. Through product adjustment and hotel scale downsizing, Wanda strives to upgrade hotel quality and build top quality hotels.

2014年万达商业规划概要

一、项目概况

2014年是继集团连续三年品质评审后的又一个关键之年。2014年万达商业地产共开业24个万达广场、18个酒店，共完成销售物业1435万平方米，累计自持物业面积达到2156.6万平方米。规划系统在保证所有项目圆满开业的同时，也确保了2014年集团的几大重点项目的高品质开业。

1.集团重点项目

前面提到的商业、酒店、文旅的3个标志性的项目一昆明“百店”、武汉瑞华、武汉“双骄”，均实现了高品质盛大开业，也都获得董事长的高度赞许。

OUTLINE OF WANDA COMMERCIAL PLANNING 2014

I. PROJECT OVERVIEW

2014 is an important year for Wanda Group after three consecutive years' quality review conducted by the Group. 2014 witnessed the opening of 24 Wanda Plazas and 18 Wanda hotels developed by Wanda Commercial Properties Co., Ltd. The area of sold real estate amounts to 14.35 million square meters and the area of accumulated self-owned real estate amounts to 21.566 million square meters. The planning system, in addition to ensuring the successful opening of all projects, also ensures the satisfactory opening of several benchmark projects of the Group in 2014.

1. HIGH-LIGHT PROJECTS OF WANDA GROUP

The above-mentioned three commercial, hotel and culture-

（图4）北京通州万达广场开业

11月29日，北京通州万达广场开业。作为长安街沿线第三座万达广场，其时尚炫酷的内外装、富于创新的空间组合及行业领先的设计运营理念，都开创了万达广场品质的全新高度（图4）。

西双版纳6星酒店及项目整体已按照集团要求顺利完成设计工作。

2.销售物业重点项目

9月6日，集团第一个“万达茂”（WANDA MALL）售楼处——南宁“万达茂”售楼处高品质开放。作为南宁、广西乃至东盟的文化、旅游、商业与生活中心，南宁“万达茂”将会影响南宁未来20 年的城市发展，而售楼处的高品质开放，则第一时间向外界展示了“万达茂”的全新形象。

3月16日“无锡展示中心”开放，12月6日“广州展示中心”开放。两个展示中心分别以“紫砂壶”和“木棉花”的造型面世；塑造建筑经典的同时，也充分诠释了当地文化内涵，成为当地民众和媒体热议的新闻热点，极大地促进了营销。无锡项目首开“日光盘”，更是在楼市整体低迷的环境下创造了行业奇迹。

全面保障总包交钥匙试点项目——四川乐山、四川德阳、佛山三水各项设计工作顺利进行。

二、绿建与安全

1.绿建标识

这一年新获97个绿建标识，其中：67个项目获得绿建设计标识，17个购物中心获得绿建运行标识，13个酒店获得绿色饭店标识。北京通州万达广场、营口项目住宅、武汉电影乐园获得“绿建二星设计认证”，大连高新万达广场获得“绿建二星运行认证”。截至2014年12月31日，万达集团累计获得“绿建标识”259项，继续位列全国企业首位。

2.节能技术

2014年，由商业规划院绿建节能研究所牵头，规划系统与商管等集团相关部门，按照《万达集团节能工作规划纲要（2011-2015年）》的要求，有计划地通过如下重点研究，进一步完善了规划系统各业态的节能工作体系：（1）《万达广场慧云系统与冷站群控系统联合模式研究》；（2）《万达广场能耗针对性解决措施研究》；（3）《万达广场能耗管控及运营监管工作报告》。

根据商管、酒管的统计数值，集团各项节能技术及管理措施均达到预期效果，均超额完成《节能纲要》中要求的2%~3%的工作目标，取得显著的节能效果。

tourism landmark projects, namely the "One Hundred Store" (the 100th plaza) in Kunming, Wuhan Wanda Reign Hotel and the "Proud Twins" in Wuhan, all welcomed grand and satisfactory openings and were highly praised by Chairman Wang Jianlin.

On 29th November, 2014, Beijing Tongzhou Wanda Plaza was opened. As the third Wanda Plaza along the Chang'an Avenue, it brings the quality of Wanda Plaza to a brand new level with its fashionable and stylish interior and exterior facades, its creative spatial combination and its leading design and operation philosophy (as shown in Fig.4).

The design of Xishuangbanna Six-star Hotel and the whole project has been successfully completed as scheduled.

2. MAJOR PROJECTS OF PROPERTIES FOR SALE

On 6th September, 2014, the first Wanda Mall sales office of the Group -- Nanning Wanda Mall sales office welcomed a high-quality opening.On 16th March, 2014, Wuxi Exhibition Center was opened, and on 6th December, Guangzhou Exhibition Center was opened.The smooth progress of the designing work of "Turnkey Contract" pilot projects in Leshan and Deyang, Sichuan Province, and Sanshui, Foshan, Guangdong Province has been fully ensured.

II. GREEN BUILDING AND SAFETY

1. GREEN BUILDING LABEL

In 2014, 97 new green building labels have been gained, wherein, 67 projects have gained green building design labels, 17 shopping centers have gained green building operation labels and 13 hotels have gained green hotel labels. Beijing Tongzhou Wanda Plaza, Yingkou Project Residence and Wuhan Movie Park have gained "Two-Star Green Building Design Label", while Dalian High-tech Wanda Plaza has gained "Two-Star Green Building Operation Label".

As of 31st December, 2014, Wanda has gained 259 "Green Labels" in total, ranking the first among all enterprises in China.

2. ENERGY-SAVING TECHNOLOGY

In 2014, led by the Green Building Energy-saving Research Institute of Wanda Commercial Planning Institute, and relevant departments of Wanda Group including the planning system, the commercial management department, etc. in accordance with the requirements in Wanda Group's *Energy Saving Planning Outline (2011-2015)*, have further improved the energy saving system for different types of business in planning system, through scheduled study of the following fields: (1) *United Model Research of Wanda Plaza's Huiyun (Intelligent-cloud) System and Cooling Station Group Control System* (2) *Research of Wanda Plaza Energy Consumption Solution* (3) *Work Report of Wanda Plaza's Energy Consumption Control and Operation Supervision*.

According to the data collected by the Commercial Management Co., Ltd. and Hotel Management Company, desired effects have been achieved as to various energy saving technologies and management measures of Wanda Group. Outperforming the requirements in *Energy Saving Outline* for 2% to 3%, it's fair to conclude that remarkable energy-saving effects have been achieved.

3. HUIYUN SYSTEM

In 2014, Huiyun system has been implemented and operated in 46 projects (including Wuhan Show Theater and Movie Park, wherein, 26 are newly built and 20 are renovated). Up to the end of 2014, the number of implemented Huiyun System added up to 50 projects in total (wherein, 28 are newly built and 22 are renovated).

3.慧云系统
全年完成46个项目（含武汉秀场和电影乐园）慧云系统上线运行（其中：新建26个、改建20个）。截至2014年底，累计共有50个项目的慧云系统上线（其中：新建28个、改建22个）。

三、科研与知识产权

2014年，商业规划系统在各产品业态中大胆创新，取得了令行业瞩目的优异成绩；共获得1个“吉尼斯世界之最”、知识产权专利9项、著作权3项、科研课题34项，填补行业和集团空白13项，荣获9项国际、国内大奖。

1.世界之最
2014年，合肥万达城展示中心荣获“大世界吉尼斯之最”之“世界最大单体鼓形建筑”。

2.知识产权
2014年，商业规划系统在知识产权工作中有重大突破，共计获得知识产权专利9项，获得著作权3项。

设计中心研发的“高层住宅安全屋”获3项“国家实用新型”专利，分别是：新型安全屋结构、安全窗组件、安全屋机电装置。这些成果在住宅产品安全设计领域属行业首创，为万达销售类物业产品在市场竞争赢得了良好口碑，通过产品核心技术极大地支持了营销。

商业规划院和商管共同研发的慧云系统获得了3项“软件著作权”（慧云智能化管理系统A版、B版、C版）；同时“慧云智能化管理系统”2014年1月已申报注册商标权（需1年公示期，2015年2月正式取得）。

3.项目获奖
全年项目获奖9项，其中：国际1项、国内8项。

商业规划院研发管控的汉街万达广场夜景照明项目在第31届“国际照明设计大会”上获照明界的“奥斯卡奖”—“IALD卓越奖”。这是中国内地第一个获此殊荣的项目。

商业规划系统首获8项“全国人居经典环境竞赛奖”，其中：商业规划院管控的3个项目首获“全国人居经典环境竞赛奖”，其中长春宽城万达广场获“建筑金奖”，上海松江万达广场景观、广州增城万达广场景观获“环境金奖”。

设计中心管控的5个项目首获“全国人居经典环境竞赛奖”，其中大连高新海公馆获“综合大奖”，东莞厚街、泉州浦西、太原龙湖3个项目获“规划、建筑”双金奖，长沙开福获“环境金奖”。

III. SCIENTIFIC RESEARCH AND INTELLECTUAL PROPERTY

In 2014, the commercial planning system has been aggressive in innovation as to various types of business, and has achieved excellent performance that stuns the counterparts, with 1 "Guinness World Record", 9 intellectual property patents, 3 licensed copyrights, 34 scientific research studies, 9 international and national awards, filling up 13 blanks of the industry and the Group.

1. WORLD RECORD
In 2014, Hefei Wanda Exhibition Center won the "Guinness Record" of "World's Largest Single Drum-Shape Architecture".

2. INTELLECTUAL PROPERTY
In 2014, the commercial planning system has achieved great breakthrough in intellectual property acquisition with 9 intellectual property patents and 3 licensed copyrights in total.

The High-rise residence safe-house developed by the Design Center won three "National Utility Model" patents, respectively being new model safe-house structure, safety window assembly and safe-house electromechanical device. These achievements, being pioneered in safety design domain of residential products, win great reputation for Wanda properties for sale in the market. Supports have been provided to sales and marketing through core product technology.

The Huiyun System, jointly developed by Wanda Commercial Planning Institute and Wanda Commercial Management Co., Ltd., won three "Software Copyrights"; meanwhile, in January 2014, declaration for trademark has been submitted as Huiyun Intelligentized Management System (after one-year publicity, the trademark be officially acquired in February 2015.)

3. PROJECT AWARDS
In the whole year of 2014, 9 projects have won awards, with 1 international award and 8 domestic awards.

The nightscape lighting project of Wuhan Han Street Wanda Plaza, developed and controlled by Wanda Commercial Planning Institute, was awarded with the "IALD Excellence Award" on the 31st Session of International Lighting Design, which is redeemed as the "Oscar of lighting industry". This is the first project in mainland China that wins the honor.

The commercial planning system has won 8 awards in National Dwelling Classic Environment Contest Prize for the first time, wherein, three projects supervised by the commercial planning institute won 3 awards in National Dwelling Classic Environment Contest Prize for the first time, in which Changchun Kuancheng Wanda Plaza won "Architecture Gold Award", and Shanghai Songjiang Wanda Plaza landscape and Guangzhou Zengcheng Wanda Plaza landscape won "Environment Gold Award".

Five projects supervised by the Design Center won 5 awards in National Dwelling Classic Environment Contest Prize for the first time, wherein, Dalian High-tech "Sea Mansion" won "Comprehensive Award", three projects respectively located in Dongguang Houjie, Quanzhou Puxi and Taiyuan Longhu won "Double Awards of Planning and Construction", and Changsha Kaifu won "Environment Gold Award".

4. SCIENTIFIC RESEARCH ACHIEVEMENTS
In 2014, the commercial planning system has completed 34 scientific research studies, including 25 of group level and 9 of system level, wherein, 10 of them fill up the blanks of Wanda Group, and the following three fill up the blanks

（图5）王健林董事长视察北京通州万达广场

4.科研标准成果

2014年商业规划系统完成科研课题34项，包括集团级课题25项、系统级课题9项。其中：填补集团空白10项，以下三项填补行业空白3项：（1）《地下四大块定额设计管控办法》（商业规划院）；（2）《超高层写字楼限额指标表》（商业规划院）；（3）《高层住宅安全屋》（设计中心）。

结语

2014年是万达商业规划非常标志性的一年。这一年，万达广场的开业数量在年终达到了109座，奠定了万达商业“百店”的基础，万达商业规划也在此基础上不断地完善和成熟。商业规划亦实现了从总图规划到施工图管控，直至后期存档的全程信息化自动管理系统，实现了以限额设计为前提的“A”、“B”、“C”三版不同造价的标准化设计，实现了广场建成后全面的“慧云商业运营管理系统”。这些成果，为集团2015年向轻资产转型，创造了成熟的技术条件，并为轻资产的“标准化”、“模块化”和“产业化”的推进提供了充足的经验。

2014年商业规划系统工作得到王健林董事长、集团丁本锡总裁、商业地产齐界总裁等领导的支持和重视。2015年万达集团将以“百店”为基，恢宏起步，进入更高速发展的轻资产时代（图5，图6）！

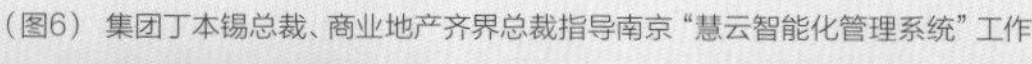
（图6）集团丁本锡总裁、商业地产齐界总裁指导南京“慧云智能化管理系统”工作

of the industry: (1) *Control Method for the Quota Design of Underground Four Sections* (by Wanda Commercial Planning Institute); (2) *Super High-rise Office Building Quota Index Table* (by Wanda Commercial Planning Institute); (3) *High-rise Residence Safety House* (by Wanda Design Center).

CONCLUSION

The year 2014 is a very important landmark year for Wanda commercial planning. By the end of this year, the number of operating Wanda Plazas has mounted to 109, laying the foundation of a "One Hundred Store" for Wanda's commercial development. In the meanwhile, Wanda's commercial planning is becoming improved and mature based on the above foundation. As for commercial planning, a complete automatic information management system from general planning to construction drawing control to filing in the later phase has been achieved; the standard design of three Wanda Plaza editions (Edition A, Edition B and Edition C) of construction costs has been achieved under the premise of quota design; an overall Huiyun Commercial Operation Management System upon the completion of plaza has been achieved. These achievements not only create mature technology conditions for the Group's asset-light transition in 2015, but also provide sufficient experiences for the advancement of asset-light's standardization, modularization and industrialization.

In 2014, the work of the commercial planning system has received support and attention from leaders including Chairman Wang Jianlin, Group CEO Ding Benxi and Commercial Properties CEO Qi Jie. In 2015, Wanda Group, based on the foundation of "One Hundred Stores", shall start with a grand momentum and enter an asset-light era with faster development speed (as shown in Fig.5 and Fig 6).

WANDA PLAZAS 万达广场

WANDA COMMERCIAL PLANNING 2014

昆明西山万达广场
东莞东城万达广场
广州增城万达广场
潍坊万达广场
上海松江万达广场
赤峰万达广场
济宁太白路万达广场
金华万达广场
常州武进万达广场
佛山南海万达广场
马鞍山万达广场
荆州万达广场

KUNMING XISHAN WANDA PLAZA
昆明西山万达广场

时间 2014 / 10 / 31 **地点** 云南 / 昆明
占地面积 7 公顷 **建筑面积** 70 万平方米

OPENED ON 31st OCTOBER, 2014
CITY KUNMING, YUNAN PROVINCE
LAND AREA 7 HECTARES **FLOOR AREA** 700,000M²

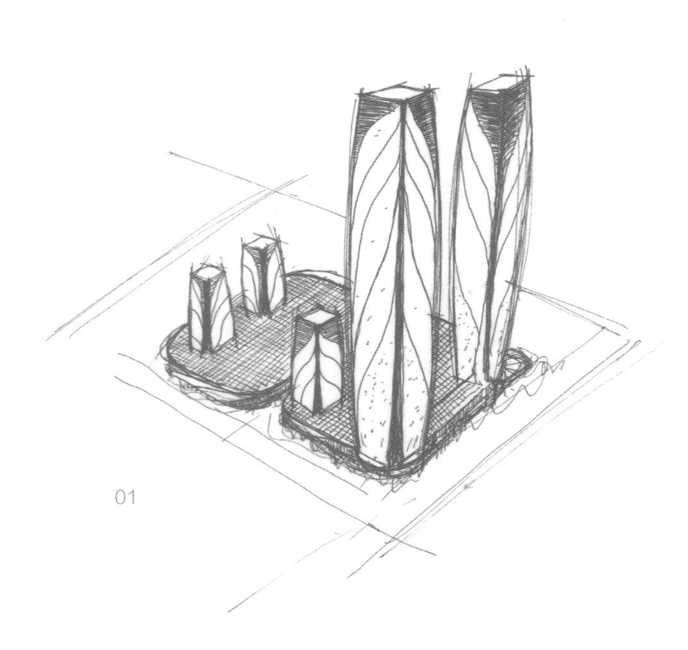

01

OVERVIEW OF PLAZA
广场概述

昆明西山万达广场地处昆明市泛亚金融产业中心园区，包括商业中心、城市步行街、超五星级酒店和超5A甲级写字楼等，占地面积7.0公顷，总建筑面积70万平方米。其中，大商业引入万达百货、沃尔玛、万达影城、国美电器、大玩家超乐城、大歌星量贩KTV等主力店，集顶级购物中心和城市酒吧街于一体；超五星级万达文华酒店和300米高“昆明双塔”甲级写字楼建成后将成为昆明金融产业中心园区乃至昆明全市的标志性建筑。

Located in the Kunming Pan-Asia Financial Industry Central Park, Kunming Xishan Wanda Plaza, with a land area of 7.0 hectares and an overall floor area of 700,000 square meters, integrates commercial center, Pedestrian Street, super five-star hotel and super 5A grade-A office building together. The large commercial area brings in quantities of anchor stores, such as Wanda Department Store, Wal-Mart, Wanda Cinema, GOME, Super Player and the Superstar, and integrates top-class shopping center and city Bar Street into one. The super five-star Wanda Vista and 300m tall grade-A office building called "Kunming Twin Tower" would be the landmark building of Kunming Pan-Asia Financial Industry Central Park and even of the Kunming city once completed.

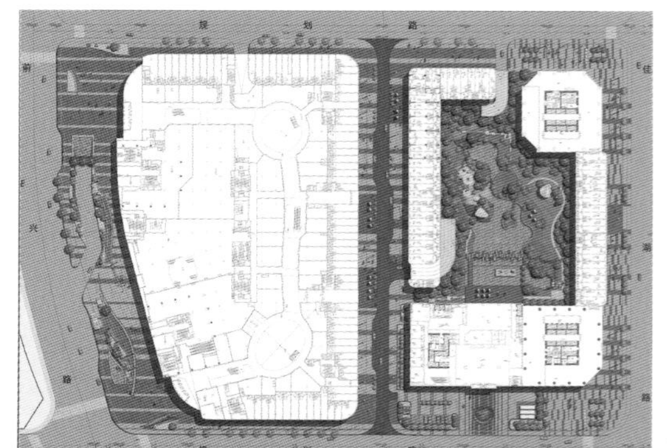

02

01 广场概念草图
02 广场总平面图
03 广场立面

03

FACADE OF PLAZA
广场外装

广场外立面方案构思来源于昆明的市花——山茶花。为突显主体雕塑的主体性，建筑立面把山茶花以简洁的线条表现出来。位于这个商区内的主体雕塑充分地考虑与建筑群相呼应的关系，成为公众与建筑之间的沟通桥梁。

塔楼由数片形似山茶花花瓣的玻璃幕墙拼合而出，体量挺拔、形态饱满，寓意为“山茶花开，欣欣向荣”；裙房以石材为主，深浅交替的色带构成舒展的幕墙水平线条，曲线的玻璃窗户凹进幕墙表面，运用虚实对比、明暗交替、色彩反差的设计手法，展现建筑柔美的线条和优雅的流畅感。

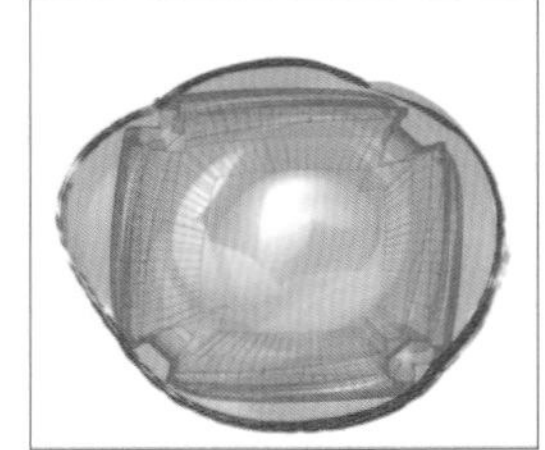
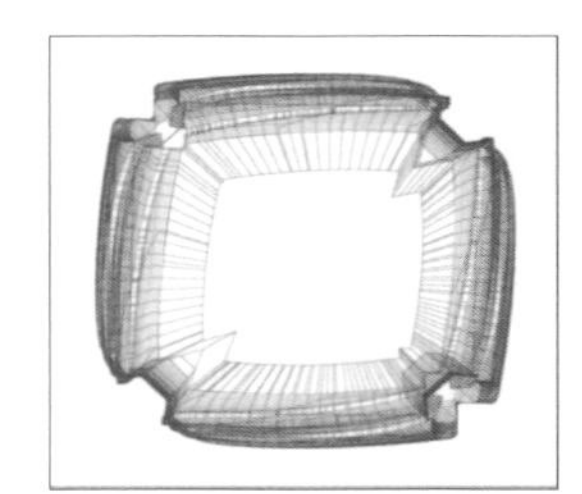

04

04 广场构思演变图
05 广场东北向外立面
06 广场北立面图

05

Conception of the plaza facade is inspired by camellia, the city flower of Kunming. To strengthen subjectivity of the major sculpture, the building facade expresses camellia in simple and clean profile lines. Major sculptures in this commercial area build a highly harmonious relationship with the building complex, setting up a communication bridge between the public space and buildings.

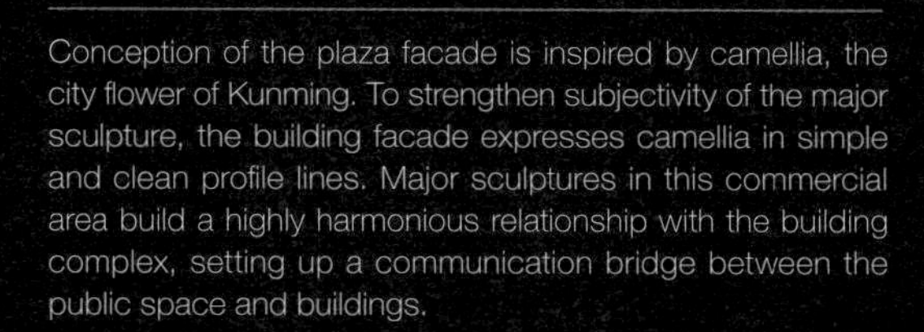

The tower is assembled by several glass curtain walls shaped like petals of the camellia, appearing to be upright in volume and plump in shape and meaning "Blooming Camellia and Growing Prosperity". The podiums, by design modes of virtual-real comparison, light-shade alternation and color contrast, take stone as the main material, utilize deep and shallow color ribbons to form smoothly extending horizontal lines of curtain walls and recess curved glass window into curtain walls, showing the soft and flexible lines and elegant sense of fluency of buildings.

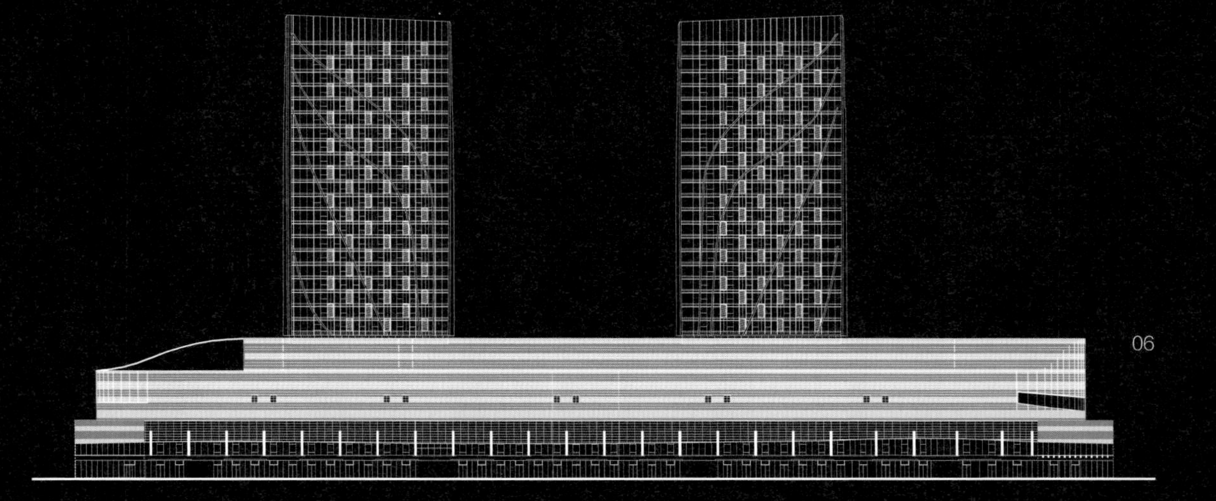

06

07

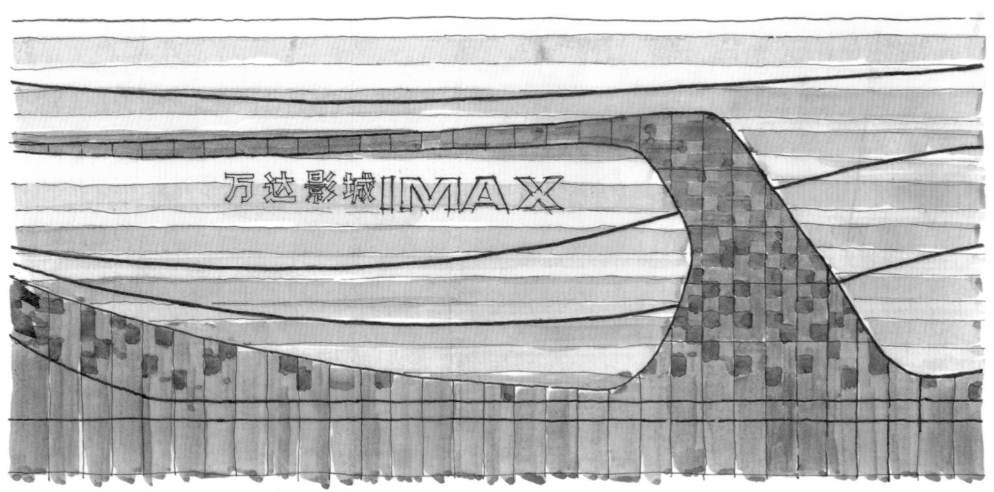

08

07 广场百货外立面
08 广场娱乐楼外立面手绘稿
09 广场南立面

09

westlink
Cabbeen
Timberland

INTERIOR OF PLAZA
广场内装

昆明西山万达广场室内空间设计与建筑设计相协调，依然采用了“山茶花”这一元素，以此作为“联系”室内外的纽带。

圆中庭为昆明西山万达广场室内设计的“亮点”，花瓣外形的观光电梯和以韵律排列的山茶花形的侧帮造型，辅以采光顶花瓣造型LED屏和几何曲线形光带，相互映衬、光影交错、虚实相间，无不体现着山茶花内敛而含蓄的优雅特性。地面采用黑白灰三种色调石材拼贴出山茶花的形态，环环相扣，层层绽放，使人仿若置身于花的世界!椭圆中庭观光电梯象征开放中的山茶花，观光电梯外形像一个柱状的花蕾，层层花瓣包裹而成，整个空间由顶面至立面到地面，使柔美的曲线贯穿整个空间。

Coordinated with the architectural design, the interior space design of Xishan Wanda Plaza continues to take camellia as the link to connect the interior with the exterior.

Highlights of interior design for Xishan Wanda Plaza fall on the circular atrium, where petal-shaped sightseeing elevator and camellia-shaped lateral wall design arranged in rhythms are lighted up by petal-shaped LED screen on the skylight and geometrically curved light bands to create a virtual and actual space filled with light and shade, fully embodying the modest and humble characteristics of camellia. Floor of the atrium is paved in the shape of camellia by stones in black, white and gray tones, sketching a recurring and blooming camellia and making people dance in the ocean of bloom. Sightseeing elevator in the oval atrium, which is designed in a pillared flower bud shape wrapped by petals, symbolizes the blooming camellia, making the whole space, from ceiling to facade and even to floor, filled with soft and flexible curves.

11

12

13

14

北入口对面设置匠心独具的立体垂直绿化墙，以开门见山的形式“叙述”昆明这座春城的特色，也充分体现了“建筑—绿化”一体化的理念。入口右侧墙面的《昆明八景》圆形浮雕体现了昆明古时的繁荣景象。

直街玻璃护栏一改传统做法，采用了无立柱玻璃栏板，侧裙板以流畅的弧线为基础、辅以山茶花造型为点缀。整个侧裙板立面以数十个山茶花造型进行组合，灵活多变。

Ingenious vertical green wall is set opposite to the north entrance to describe features of Kunming, the City of Perpetual Spring, in a direct and straight way. This design also embodies the conception of integrated “Building and Green”. The “Eight Sights in Kunming” medallion on walls on the right side of the entrance represents the prosperity of Kunming in ancient times.

Glass railings for straight gallery spaces, shifting from conventional practice, take glass balusters without uprights instead. Side skirt boards are designed by flowing curves and embellished by camellia shapes. Dozens of diversified camellia shapes are combined for facade design of the whole side skirt board, flexible and changeable.

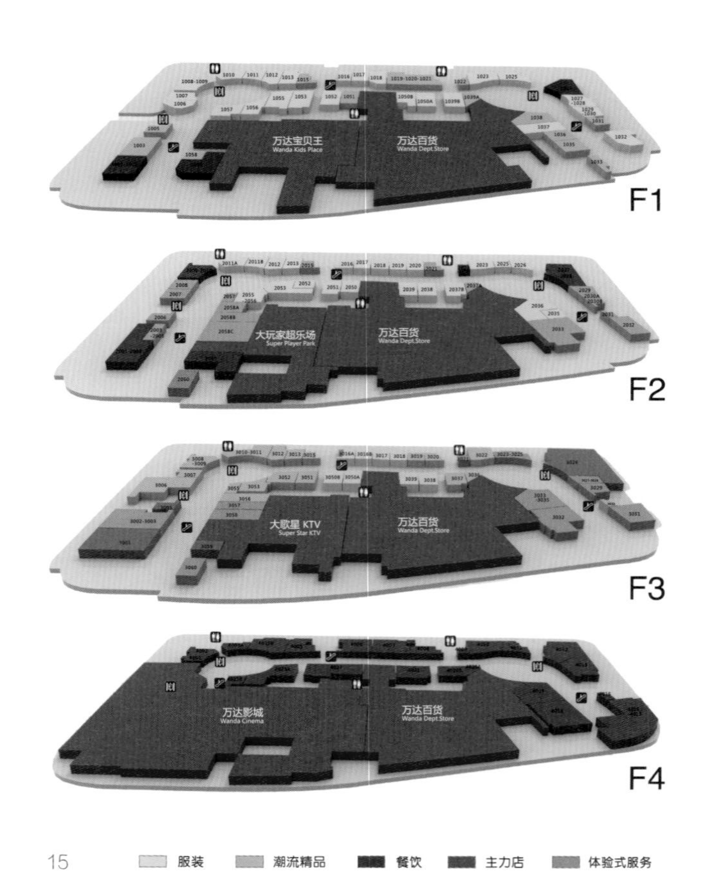

15

13 植物墙
14 北入口
15 商铺落位图
16 连桥

16

WANDA DEPT. STORE
万达百货

昆明西山万达广场的万达百货定位为精致生活店，于2014年10月31日正式开业，建筑面积近3.0万平方米，使用面积达1.8万平方米，有近200个品牌。其中一层为“国际精品馆”（经营国际精品、黄金珠宝、化妆品、女鞋），目标客群为20~40岁时尚中高端客群；二层为“名媛风尚馆”（经营淑女装、女内衣、女包、女饰品），目标客群为20~45岁女性客群；三层为“时尚潮流馆”（经营轻淑女装、轻淑女饰品），目标客群为18~30岁时尚女性客群；四层为“绅士休闲馆”（经营男正装、男休闲装、户外用品、男箱包、男配饰），目标客群为20~45岁中高端男性客群；五层为“童用生活馆”（装童装、儿童玩具、小家电、床品），目标客群为18~35岁年轻家长。

Wanda department store set in Xishan Wanda Plaza, opened on 31st October, 2014, is positioned as an exquisite life store and covers a floor area around 30,000 square meters and net area of 18,000 square meters. About 200 brands join the Plaza, of which the ground floor is arranged as the International Boutiques that deals with internationally competitive products, gold & jewelry, cosmetics and lady's footwear and targets at fashionable middle and high-end clients between ages of 20~40; the first floor is positioned as the Ladies Fashion that sells ladies clothes, underwear, bag and accessories for 20~45 years-old ladies; the second floor is the Fashion Trends engaged in young lady fitted and accessories for 18~30 years-old stylish female customers; the third floor is the Gentleman Wear that deals with men's formal dresses, casual wear, outdoor goods, bags and suitcases and accessories for 20~45 years-old middle and high-end male customers; the fourth floor is the Children Products that sells children's garments, toys, small appliances and bedding and aims at young parents between 18~35 years old.

17 万达百货走廊
18 万达百货中庭
19 万达百货店面

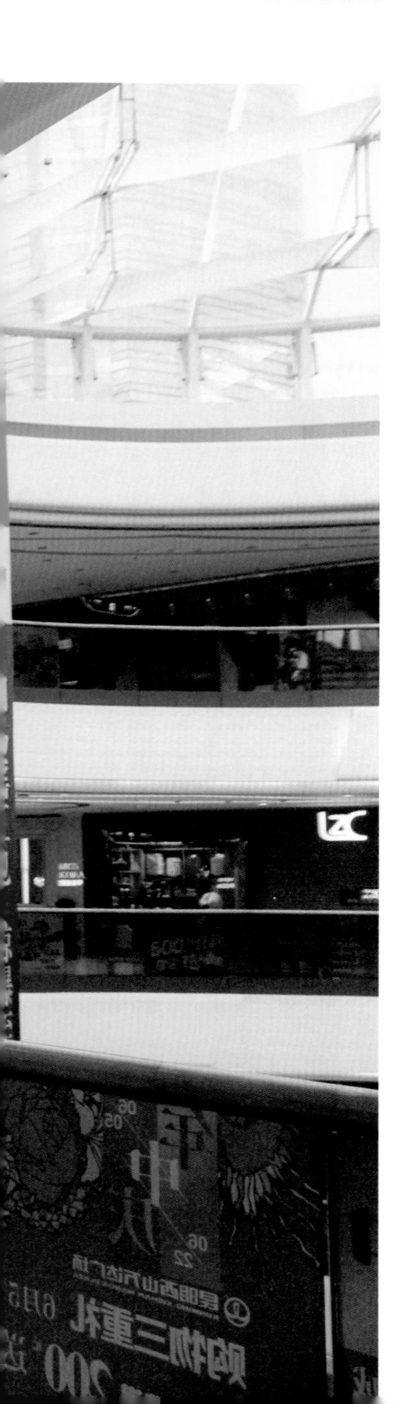

18

19

昆明西山万达影城位于万达广场四楼，拥有15个厅（包括1个IMAX厅、两个VIP厅、两个2D厅和10个3D厅），是云南省最大的五星级影城，可容纳近2100人同时观影。作为万达院线在云南最大、最豪华的影城，拥有当今世界顶级的IMAX放映系统、TMS数字放映管控平台、杜比7.1声道音频解析技术等配置，以及先进的网络自助取票系统。影城设计突出云南当地独有特点，营造具有独特氛围的观影空间：（1）以开屏孔雀的“尾雉翎”为主题提炼设计元素，系统运用到影城大堂、走廊等区域的天花造型上；（2）结合多民族文化特点，将代表性的人物、服饰、名称等符号化加以运用；（3）将云南电影历史上代表性的电影，通过“电影照片墙”、“拉布灯箱”等形式进行展示。

As the largest five-star cinema in Yunnan province, Xishan Wanda Cinema has 15 halls including one IMAX, two VIP halls, two 2D halls and ten 3D halls and can accommodate total 2100 persons. In the largest and most luxurious cinema among Wanda cinema line in Yunnan province, its IMAX enjoys the top projection system of the world, TMS digital projection control platform and Dolby 7.1 channel audio resolution technology as well as advanced online ticket self-service system. Design of the cinema highlights local features of Yunnan province and seeks to create a unique viewing space through the following efforts. Firstly, "Tail Feathers" of a peacock in his pride is set as the theme in design elements extraction and systematically applied in ceiling modeling of lobby and corridor of the cinema. Secondly, characteristics of multi-ethnic culture, such as representative figures, clothes, designations and other symbolizations, are integrated into the design. Thirdly, representative movies in the film history of Yunnan Province are presented through movie photo wall, fabric light box, etc.

22

20 万达影城入口
21 万达影城 VIP 厅
22 万达影城走廊

23

THE SUPERSTAR
大歌星KTV

昆明西山万达大歌星位于广场三层，于2014年10月31日开业，内设有总统包、VIP包房、商务包房等各类大中小包房。大歌星落户昆明，为当地市民提供一个尽情欢唱的魅力舞台。大歌星运用万达总部量身定做的点歌系统，点歌界面快捷方便，可实现拼音、歌星、语种等多种点歌方式，超大曲库，定期更新，满足消费者对曲库的多样性需求。普通包厢影音设备内置多种KTV演唱效果，搭配无线话筒及顶级音响，使声音纯净稳定，演唱更加轻松。总统包厢搭载世界顶级品牌设备，音色饱满，穿透力强；并且配有专业音频处理设备，可以体验震撼的效果。

The Superstar, located on the second floor of the Plaza, is opened on 31st October, 2014 and furnished with boxes in various standards such as the president, VIP and business boxes, etc. Settling of the Superstar in Kunming Province sets a charming stage for the public to enjoy singing. The Superstar utilizes VOD system tailored by the Wanda headquarter that enjoys quick and easy operation interface and enables operation by phonetic Chinese alphabet, singer or language, moreover, the song list is periodically updated to satisfy customers' requirements on diversity of songs. AV equipment for general boxes which are inbuilt with multi-KTV performance effects, together with wireless microphone and top-class stereo system, guarantee a pure and stable voice for easier singing; the president box accommodates top-class brand equipment of the world with advantages of full range of tones and powerful sounds, and professional AV processing equipment, giving customers an opportunity to experience the shocking effect.

23 大歌星 VIP 包房
24 大歌星大厅
25 大歌星总统包房
26 大歌星主题包房

24

25

26

27 万达宝贝王
28 万达宝贝王
29 万达宝贝王
30 万达宝贝王
31 万达宝贝王入口
32 万达宝贝王

31

32

WANDA KIDS PLACE
万达宝贝王

昆明西山万达儿童天地，面积3500平方米，是万达专为2~8岁中国亲子家庭设计、首期推出的动漫亲子乐园之一。儿童天地通过奇趣动漫主题氛围的营造，在精彩纷呈的游乐中融入丰富多彩的体验方式，旨在打造一个拓展儿童思维、培养儿童社交、丰富儿童生活、增进亲子情感的欢乐世界；设置了多种有趣的设施或区域，如迷你爬山车、旋转飞椅、夺宝奇兵、益智游戏互动不停、亲子美食童趣多多等。

Xishan Wanda Kids Place, with an area of 3,500 square meters, is one of cartoon family parks designed for Chinese children between ages of two to eight and firstly introduced by Wanda. It, through creating a quaint cartoon themed atmosphere and integrating varieties of experience modes into wonderful entertainment, aims to build a happy world where children's thinking would be expanded, social skills be cultivated, living be enriched and parent-kid relationship would also be improved. Here a large quantity of interesting facilities and areas are set, such as Mini mountain bike, flying chair, Indiana Jones, puzzle park, family food & beverages, etc.

33

LANDSCAPE OF PLAZA
广场景观

建筑的主题是“山茶花”。景观契合建筑的主题，使景观与建筑形成内在的协调。景观采用“生长的叶”的主题，通过立体三维景观的塑造，传递“花与叶”的关系；“生长的叶”也体现在地面铺装、花坛、树池以及景观小品与植物上。通过多层次的立体打造，“花与叶”呈现出全方位的再现。

"Camellia" is the theme of the architecture. Coincide well with this theme, the landscape design reaches an internal coordination with architecture, adopts "Growing Leaf" as the theme and describes the relationship between "Flower and Leaf" through setting of three-dimensional landscape. Through every detail like the floor pavement, parterre, planters, featured landscape and plants, the landscape theme is outstanding, and through such arrangement in multi levels, "Flower and Leaf" relation is fully represented.

34

35

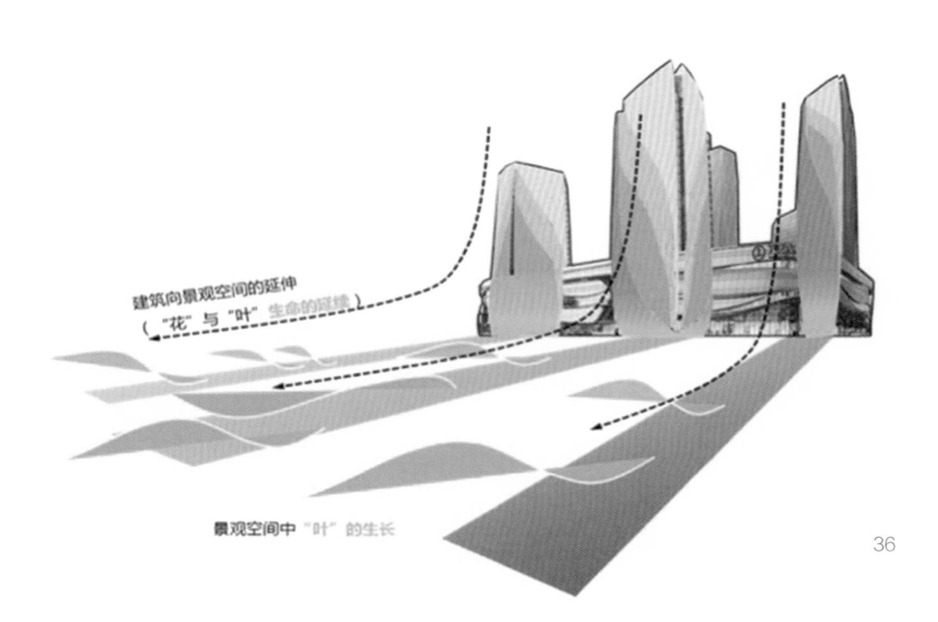

36

NIGHTSCAPE OF PLAZA
广场夜景

夜景塑造时始终把用光刻画“七彩云南”放在核心地位，在颜色上以“日照金山的赤金”、“山茶花的粉黛”、“孔雀羽毛的靛蓝”、“红河谷的赭红”和“蜡染的翠绿”等标志性颜色作为动画的主色调，画面绚丽但不失稳重，突出了其显赫地位。夜景照明动画的设计上将当地特色的地域文化符号提炼升华，形成“七彩云南”、“山茶花语”、“滇川风情”和“星光火舞”四大主题、十个场景，成为一个完整的体系。该项目夜景照明设计获2014爱迪生照明亚洲金奖。

Nightscape of the Xishan Wanda Plaza always adheres to sketching "Colorful Yunnan" by light. In terms of color, many symbolic colors, such as "Gold like the Meri Snow Mountain at Sunrise", "Pink like the Camellia", "Indigo like Peacock Feathers", "Red Ocher like the Red River Valley" and "Green like Batik Painting", are used as bases of the animation to draw a bright yet steady picture, in an attempt to strengthen "Colorful Yunnan". As for animation design of the Plaza, regional culture symbols with local characteristics of Yunnan are refined and sublimated to create four themes titled "Colorful Yunnan", "Camellia Language", "Yunnan & Sichuan Scenery" and "Starlight Dance" and ten scenes, forming a complete system. Nightscape lighting of Kunming Xishan Wanda Plaza was awarded with "Asia-Pacific Lighting Golden Medal" of "Edison Award 2014".

37

33 景观主题雕塑——生长的叶
34 景观小品——麒麟
35 景观小品——凤凰
36 景观设计意象图
37 广场夜景

38

EXTERIOR PEDESTRIAN STREET
室外步行街

室外步行街立面寓意“七彩云南”，色彩丰富而绚烂，融合近代法式建筑风格、民居建筑风格和现代主义建筑风格，体现昆明地方建筑的历史沿革和多民族共融的特点。立面材料通过运用石材、灰砖、红砖、彩色玻璃、金属构架的错落搭配，营造出丰富多彩的商业氛围。在局部主题店铺设计中使用云南当地的文化符号，以突出地域风情。

Facade of the exterior pedestrian street also symbolizes the "Colorful Yunnan". The rich and exuberant colors and the integration of modern French architectural style, dwelling architectural style and modernism architectural style reflect historical development and multinational fusion of local architecture in Yunnan. Well-arranged stone, grey brick, red brick, stained glass and metal framework enrich the facade materials and build a colorful commercial atmosphere. Local culture symbols of Yunnan are applied in the design of some theme stores to highlight regional customs.

38 室外步行街店面
39 室外步行街设计手绘稿
40 室外步行街
41 室外步行街入口
42 室外步行街外立面

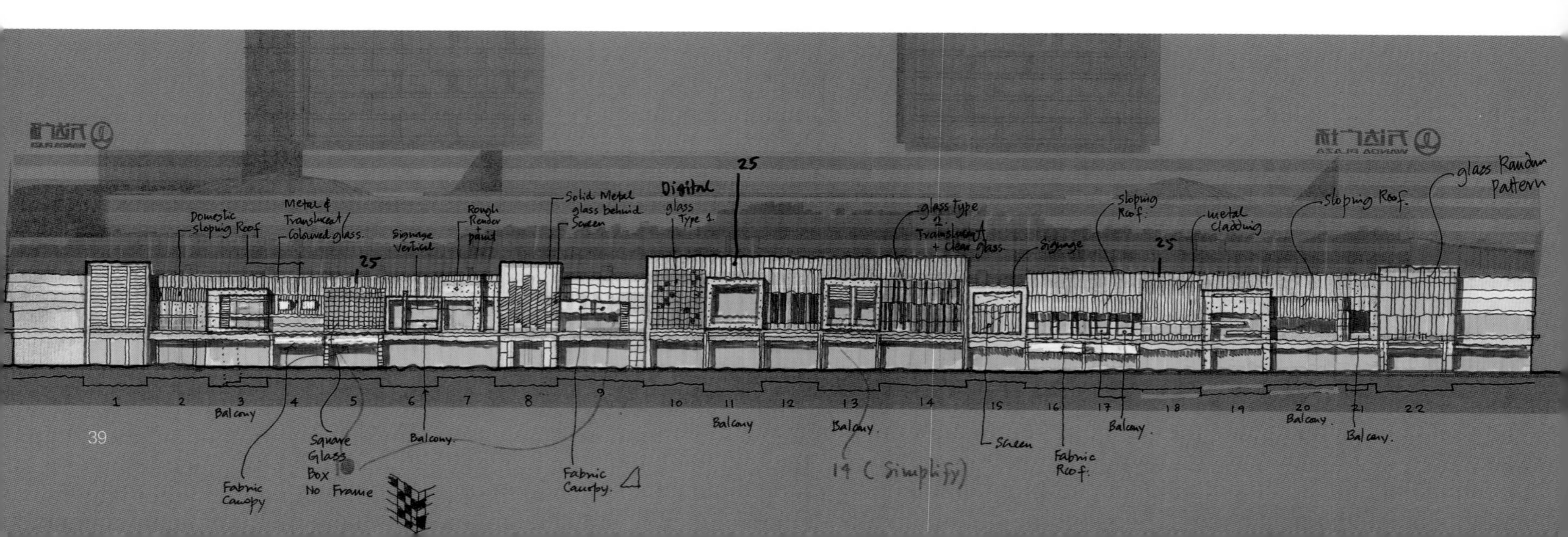

39

40

41

42

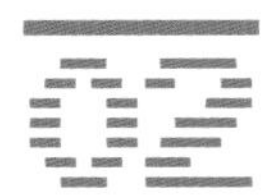

DONGGUAN DONGCHENG WANDA PLAZA
东莞东城万达广场

时间 2014 / 09 / 12 **地点** 广东 / 东莞
占地面积 12.39 公顷 **建筑面积** 53.24 万平方米

OPENED ON 12th SEPTEMBER, 2014
LOCATION DONGGUAN, GUANGDONG PROVINCE
LAND AREA 12.39 HECTARES
FLOOR AREA 532,400M²

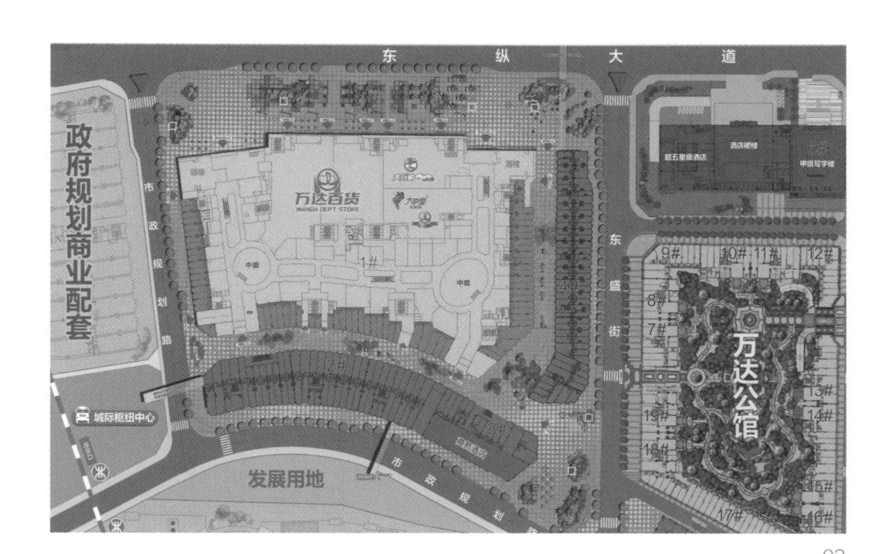

02

01 广场鸟瞰
02 广场总平面图

OVERVIEW OF PLAZA
广场概述

东莞东城万达广场位于东莞市东城区东城中路与东纵路的交汇处，属于东莞城区目前最繁华的商圈，商业氛围浓厚，交通便捷。占地约12公顷，总建筑面积约53万平方米，其中地上建筑面积40万平方米，地下建筑面积13万平方米。广场含购物中心、六星级酒店、甲级写字楼、公寓（写字楼）、室外商业街等，其中，购物中心引入万达百货、宝贝王、万达影城、大玩家、大歌星量贩KTV等大型主力店。东莞东城万达广场成为东莞最具代表的城市综合体，是整个东莞城区的商业配套，辐射珠三角及港澳台，更是城市的产业集群，极大地改变目前东莞商业布局，创造大量就业岗位及税收，成为东莞新的城市地标、新的城市中心。

Dongguan Dongcheng Wanda Plaza is located at the intersection of Dongcheng Middle Road and East Longitudinal Road in Dongcheng District of Dongguan, the most prosperous business district in urban area of Dongguan, and enjoys well commercial atmosphere and convenient transportation. With a land occupancy of 12 hectares and gross floor area of 530,000 square meters, including aboveground floor area of 400,000 square meters and underground floor area of 130,000 square meters, the Plaza integrates shopping center, 6-star hotel, grade-A office building, apartment (office building) and exterior commercial street together, of which the shopping center attracts Wanda Department Store, Kids Place, Wanda Cinema, Super Player, Superstar and other anchor stores. The Plaza is the most representative urban complex in Dongguan and served as the commercial facilities for the whole city radiating the Pearl River Delta and HMT area, and even the industrial cluster of Gongguan that improves commercial layout of the city and creates plenty of jobs and revenues, becoming the new urban landmark and city center of Dongguan.

03

03 广场外立面
04 广场外立面结构分解图
05 广场主入口

FACADE OF PLAZA
广场外装

在简洁挺拔的亮银铝幕墙上，建筑入口、幕墙玻璃盒子、购物中心雨篷采用相同的红棕色彩釉玻璃，用现代手法诠释中国古典灯笼意向，半透明的质感在天光、灯光下幻化出四时不同的多变色彩。

建筑设计通过精炼的材质、组合的形体、变换的色彩和肌理形成简洁、大气的购物中心立面效果。利用建筑表皮的连续性与不同尺度及功能的盒子体量的结合，形成既风格统一又层次丰富的商业立面效果。

On the background of simple, stiff but smooth bright silver aluminum curtain wall, the same reddish brown enamelled glass are adopted at the building entrance, glassy box of curtain wall and canopy of shopping center to express the meaning of Chinese traditional lantern in a modern way. The translucent texture will produce changing colors at different times under the daylight and lights.

Architectural design of the Plaza, through refined materials, combined shapes and changing colors and textures, creates a simple but grand shopping center facade effect. Continuity of the building surface and combination of box volumes with different sizes and functions help to form a commercial facade effect in unified style but with rich layers.

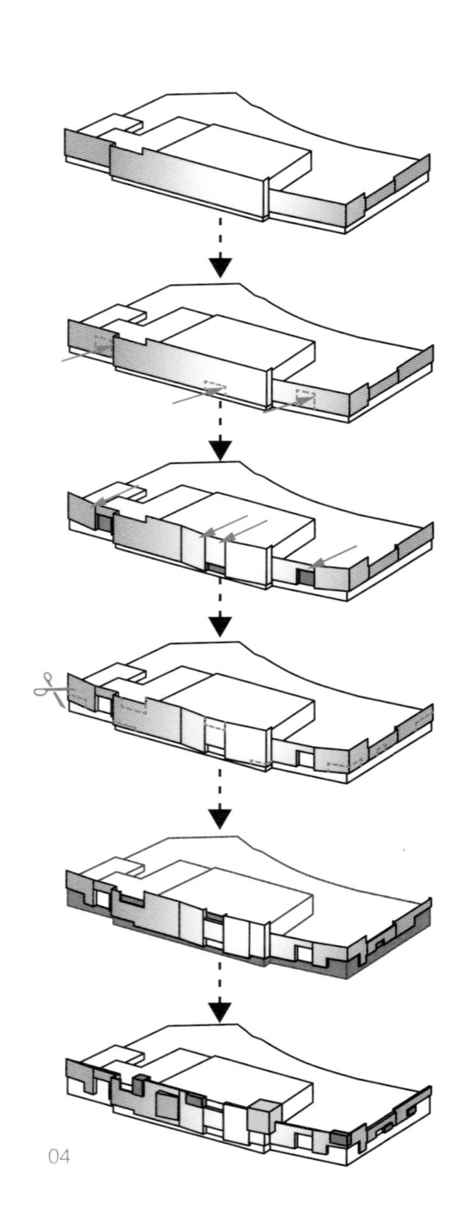

04

05

建筑表皮的处理采用模数化系统，同一主题和肌理的金属穿孔板仿佛建筑的皮肤般，匀质分布于建筑表面，整合并重新定义了建筑的体量。

建筑立面上，两个方型广告灯箱成为主要的视觉焦点，加之其体量及位置，极好地定义了建筑的商业性质并营造出很好的商业氛围。巨型玻璃广告位采用拉索式幕墙，确保通透效果也便于后期更换维护。

Benefiting from modular system, treatment of the building surface achieves an extraordinary effect. Just like skins, metal perforated plates in the same theme and texture are evenly distributed on the building surface, integrating and redefining volume of the building.

On the building facade, two square advertising boards, combined with its volume and position, become the main visual focus and define commercial nature of the building extremely well and also create an appealing commercial atmosphere. The giant glass AD adopts cable-truss type curtain wall to ensure the transparent effect and convenient future replacement and maintenance.

06 广场百货外立面

1F

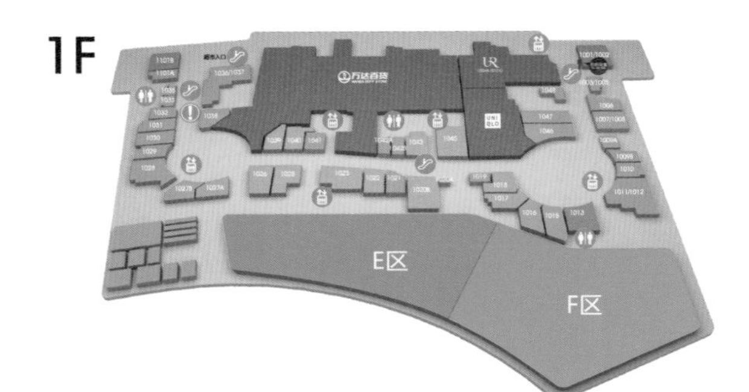

2F

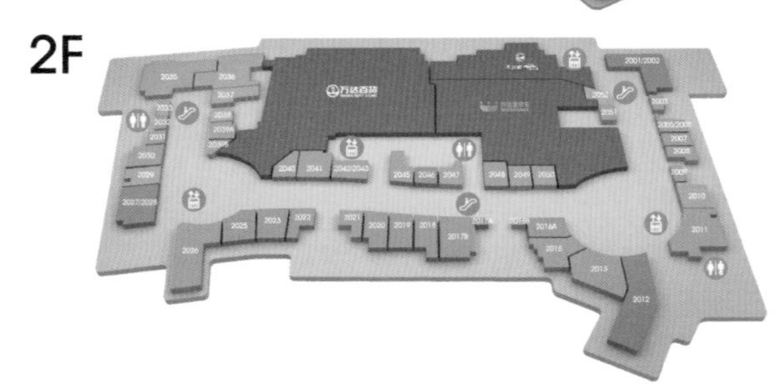

3F

服装　餐饮　精品　体验

07

07 商铺落位图
08 圆中庭
09 椭圆中庭地面拼花
10 椭圆中庭

08

09

10

INTERIOR OF PLAZA
广场内装

项目内装设计以“圆舞青春，环动时尚”为设计主题。“圆”是最能展现“中心”概念的符号，它也是最为完美而和谐的形状。圆也与外立面建筑语言有着紧密的联系，并通过圆形的汇聚与分散，从视觉上标示出不同空间的层次地位，体现出明晰的交通节点。

Interior of the Plaza is designed by focusing on the theme of “Round for Youth, Ring for Fashion”. Round is the symbol that can best represent the concept of “center” and the most perfect and harmonious shape. Moreover, round is closely linked with architectural language of the exterior facade. Gathering and scattering of round shape visually indicates layer and gradation of different spaces and show distinct traffic nodes.

11

室内步行街中庭采光顶里，“圆”化身于一个个小孔，镶嵌在顶上的铝板形成的体块里，既起到了部分遮阳的功能，还产生了丰富的视觉光影。

As for the interior pedestrian street, "round" is expressed in lots of small holes inlaid in blocks formed by top aluminum plates, not only working for sun-shading, and also creating rich visual light and shadow effect.

11 扶梯
12 直街

12

13

14

SUB-ANCHOR AND SPECIALTY CATERING
次主力店及特色餐饮

东莞东城万达广场是万达集团在华南地区正式开业的首家旗舰店，不仅筑造了一个城市新中心，更是把万达广场的品质再次推向了一个崭新的高度。相对于广场外立面装修时尚大气、公共区装修高端简约，商业部分品牌组合强、体验业态丰富、整体结构流畅；尤其是“一店一色”装修呈现侧立招牌的设计创新，次主力店UR、优衣库设计简约大气的风格，及具有酒吧风格的餐饮装修等特色，使其成为2014年新开业万达广场中“一店一色”的装修典范。

Dongguan Dongcheng Wanda Plaza is the first flagship store formally opened by Wanda Group in south China, upon which a new city center is established hereof and quality of Wanda Plaza has also been raised to a new level. Finish of the public area, compared to fashionable and grand finish of the exterior face, appears to be simple but high-end. Commercially, effective combination of brands, abundant experience activities and well-organized structure are also highlights of the design. A design innovation demonstrated by side-mounted shop signs in "one shop, one style" decoration, simple but grand design style of UR, Uniqlo and other sub-anchors, and finish of caterings in bar style promotes Dongguan Dongcheng Wanda Plaza to be a design paradigm of "one shop, one style" for Wanda plazas newly opened in year 2014.

15

16

LANDSCAPE OF PLAZA
广场景观

广场通过高大乔木树阵凸显主入口位置，同时发挥标识作用，指引人流，提高业态动线的自然感。在植物选择中运用大王椰子、银海枣、凤凰木等植物的高大特征，体现仪式感与庄重氛围。

Tall tree array is designed for the Plaza to stress position of the main entrance, guide people as a marker and improve naturalness of the industry flow. Roystonea regia, Phoenix sylvestris Roxb and Delonix regia, benefiting from their tall characteristic, are planted to create a ceremony-like solemn atmosphere.

13 特色餐饮店面
14 特色餐饮店面
15 次主力店店面
16 景观主题雕塑

17

18

19

17 广场水景
18 景观小品
19 广场景观灯
20 景观绿化

20

21

22

NIGHTSCAPE OF PLAZA
广场夜景

项目夜景设计通过灯光把穿孔铝板、彩釉玻璃等不同体块有机组合形成统一，形成了有机的光影变化和明暗特色，让夜晚的灯光可静谧而富有诗意，可大气磅礴而光影涌动。

Nightscape design of the Plaza, through organic combination of perforated aluminum plate, enamelled glass and other different blocks, forms a unified effect. Organic light and shadow thus presented make the light to be quite and poetic or to be magnificent and surging as night falls.

EXTERIOR PEDESTRIAN STREET 室外步行街

室外步行街流线与城市主要人流动线有机接驳，与大商业人流内外交融。立面采用与大商业相同的立面设计元素，通过精致细腻的现代设计风格、丰富的形体处理、富于空间变化的连廊系统、通透多彩的玻璃广告位、比例精确的侧招和通长的店招、精心设计的景观元素等处理，使得室外步行街无论是日景还是夜景均散发着浓郁的商业气息。

Exterior pedestrian street streamline of the Plaza is organically connected with the main people flow of the city and completely integrated with people flow of large commercial area. The facade design elements adopted for large commercial center are continuously applied in facade design of the exterior pedestrian street. The delicate and exquisite modern design style, diversified shape processing, corridor system with varied spatial effects, transparent and colorful glass ad, side-mounted signboards in exact proportion and full-length signage, well-designed landscape elements and other similar details all endow the exterior pedestrian street with rich commercial atmosphere, day or night.

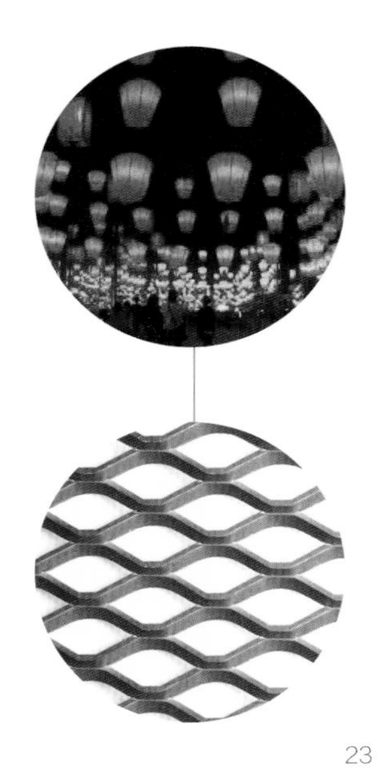

23

25

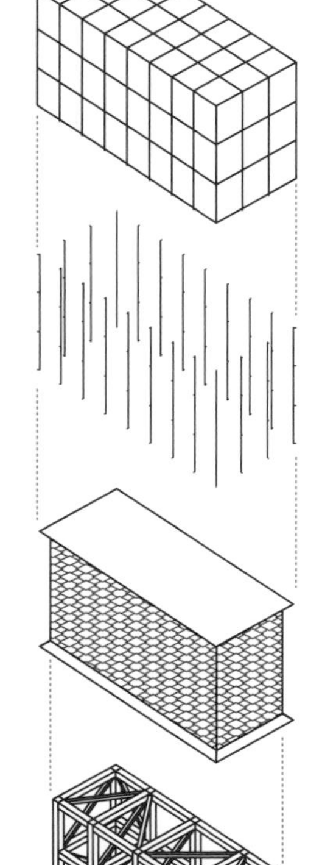

24

21 景观主题雕塑
22 广场夜景
23 室外步行街设计意象图
24 室外步行街体块分析图
25 室外步行街景观
26 室外步行街

26

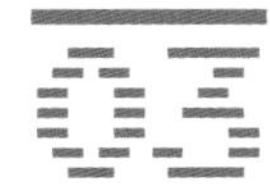

GUANGZHOU ZENGCHENG WANDA PLAZA
广州增城万达广场

时间 2014 / 05 / 16　**地点** 广东 / 广州
占地面积 8.95 公顷　**建筑面积** 38.25 万平方米

OPENED ON 16th MAY, 2014
LOCATION GUANGDONG / GUANGZHOU
LAND AREA 8.95 HECTARES　**FLOOR AREA** 382,500M²

OVERVIEW OF PLAZA
广场概述

广州增城万达广场位于增城市中心地带，占地8.95公顷，总建筑面积38.25万平方米，是集购物中心、五星级酒店、商务酒店、SOHO、室外步行街等几大功能业态于一体的大型商业项目。

Guangzhou Zengcheng Wanda Plaza is located in the downtown of Zengcheng and covers a land area of 8.95 hectares and gross floor area of 382,500 square meters. The Plaza is a large commercial project that integrates several functional businesses including shopping center, 5-star hotel, business hotel, SOHO and exterior pedestrian street into one.

01 广场鸟瞰
02 广场总平面图

02

03

04

03 广场主入口
04 广场百货外立面
05 广场外立面

05

FACADE OF PLAZA
广场外装

整体建筑体量根据周边环境进行了有机切割，融入增城特有的历史文化，形成了5个大小形态各异的体块，同时将它们和谐地组织在一起，与周边城市环境融为一体。大商业建筑左右两个门头采用门式构图，在此基础上进行了弯曲和切割变形，下小上大，形成极富夸张的造型，并运用玻璃等材质营造出水晶门的感觉，晶莹闪耀，同时产生波光粼粼的效果，与增城牌楼和增江水文化吻合；而紧贴门头表皮的铝板线条也衬托出门头造型多端变化。中间左侧体块为一个文化宝盒概念设计，内为镂空艺术穿孔铝板，外绕石材彩带，寓意来自何仙姑的凤舞之姿，引入增城的文化概念，并将其抽象地表达在建筑设计当中。

The whole building volume, combing surrounding environment of the building and unique history and culture of Zengcheng city, is organically divided into five blocks in different sizes and shapes. Meanwhile, these blocks are harmoniously organized together to integrate with surrounding urban environment. The two gates on both sides of building of large commercial area adopts gate-type structure with additional bending and cutting deformation to form an extremely exaggerative shape, large on top and narrow below. Besides, glass and similar materials are adopted to build a crystal gate, bright and shining, and create a sparkling effect, echoing archway and water culture of Zengcheng city; lines of aluminum plates close to the gate further highlight the changeful gate shape. Among the five blocks, inspired by graceful dancing postures of Immortal Woman He, the middle left one is designed in the concept of culture box with hollow-out perforated aluminum plate inside and stone color belts surrounded. Culture symbol of Zengcheng city is hereby introduced into the design and abstractly expressed in the architectural design.

06

06 椭圆中庭
07 商铺落位图
08 连桥

INTERIOR OF PLAZA
广场内装

本案的主题及设计元素均以增城之名的得来为依据，演变成琼台叠玉的空间形态和线型层叠与变化的元素符号。“明日出天山，苍茫云海间”。远处的一瞥红霞、侧帮的线条勾勒出的云彩，如站在云海山峰中，享受惬意——这就是长街所要表达的意境。线条勾勒出的侧帮，犹如有行云流水；交错的廊桥丰富了空间，蓝色玻璃犹如日出之前那片彩霞，似有非有。整体长街通过白色GRG、艺术玻璃及不锈钢营造了一个高端的购物中心形象。

Origin of the city name, Zengcheng lays a foundation for theme and elements of interior design of the Plaza and evolves into spatial form of jade platform and symbol of changing lines. As the poems goes, "Rising from the Tianshan Mountains, Bright Moon Shines in the Boundless Sea of Clouds", a trace of red glow afar and clouds outlined by lateral walls make customers enjoy and experience the pleasure like standing on a mountain hidden in the sea of clouds. This is exactly the artistic conception that the pedestrian street intends to express. Lateral walls outlined by lines are just like floating clouds and flowing water, winding gallery bridges enrich the spatial level and the blue glass is just like the trace of rosy clouds before sunrise, existing but indistinct. The pedestrian street, by means of white GRG, art glass and stainless steel, builds up the image of a high-end shopping center.

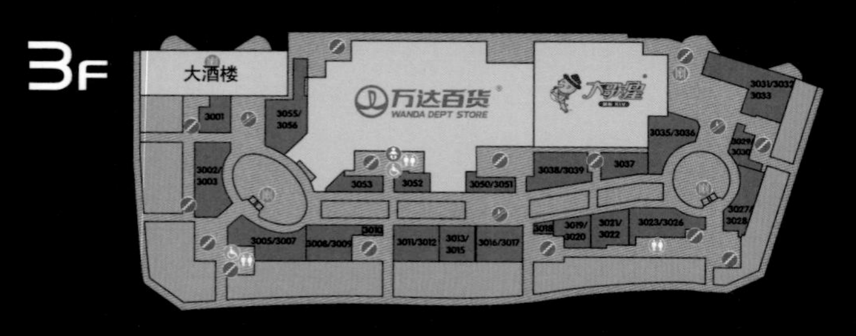

07

08

09

THE SUPERSTAR
大歌星KTV

大歌星（量贩KTV）位于广场的3楼，于2014年5月16日开业，是全国第82家门店，营业面积3500平方米，日均人流量占广场日均人流量的8%。大歌星共有55间包厢（其中量贩区48间，VIP区7间），设有总统包房、VIP包房、商务包房、主题包房和中小包房，能满足顾客的不同需要。多样的装修风格为顾客提供丰富的体验，设有主题包房——星座密语主题包房、淑女世界主题包房、篮球宝贝主题包房、足球宝贝主题包房及幻影世界主题包房等。

The Superstar (KTV), located on the second floor of Guangzhou Zengcheng Wanda Plaza and opened on 16th May, 2014, is the 82nd store all over China, covers a business area of 3,500 square meters and occupies 8% average daily visitors flow rate of the Plaza. The Superstar is furnished with 55 boxes (7 in VIP area and 48 in the rest) including president, VIP, business, theme and medium-sized and mini boxes to satisfy different demands of customers. The diversified finishing style here provide customers with a variety of experiences, such as the various theme boxes including Constellation Code, Ladies World, Basketball Baby, Football Baby, Phantom World, etc.

10

LANDSCAPE OF PLAZA
广场景观

景观设计以岭南山水为概念设计主题，尝试将传统的观赏性商业景观空间转向以体验互动为主的功能性景观场所。项目铺装粤剧水袖舞元素提炼出的大曲线贯穿始终，统筹整个景观秩序，使景观与建筑立面有机融合。考虑带形广场将近60米的面宽，采用突破常规的景观空间处理手法，利用水系作为两个广场的纽带。其间，疏林草坡、灯柱广场、木平台、水岸大台阶以及景观桥相互交织在一起，使整个景观空间更具灵动性和趣味性，重点增加了空间体验性和互动性。

Landscape design of the Plaza takes Lingnan's landscape as the conceptual design theme and attempts to change traditional ornamental commercial landscape space into an experience and interaction-oriented functional landscape site. Pavement of big curves refined from elements of long sleeves works of Cantonese opera is running throughout the whole design to plan the whole landscape order, achieving organic integration of landscape and building facade. In view of nearly 60m width of the Plaza, unconventional landscape space design is adopted, in which water system is utilized as the link between two squares. Scarce forest and grassland, lamp pole square, timber platform, waterfront step and landscape bridges are intertwined together to make the whole landscape space more flexible and interesting and enhance the experience and interaction.

09 大歌星商务包房
10 大歌星总统包房
11 广场景观水景

11

NIGHTSCAPE OF PLAZA
广场夜景

14

12 广场景观
13 广场夜景
14 室外步行街
15 景观小品
16 室外步行街
17 室外步行街

15

16

17

EXTERIOR PEDESTRIAN STREET
室外步行街

室外步行街有六大特点：(1)采用中西结合的广州骑楼立面造型，如简化的方形柱式、柱础及立面几何装饰图案。(2)运用镬耳屋顶和马头山墙，展现岭南建筑特色。(3)外街模块细部运用岭南建筑材料，立面装饰花窗及出挑雨篷，丰富模块造型。(4)欧式建筑细部造型融入外街模块，如同岭南文化中吸收的西方文化。(5)通过对不同的设计模块的组合、连接形成一体，同时也展现出岭南文化的独特魅力。(6)景观从粤剧水袖舞为主题，起伏的花坛、雕塑小品、旱喷等在大曲线肌理中有机布置。

The exterior pedestrian street has the following six characteristics: first, it adopts the facade modeling of Guangzhou Qilou (special local structure) integrating both Chinese and Western features, and applies simplified square column type, column base, and facade geometric decorative pattern; second, it adopts pot handle-shaped roof and crow gable to reveal the architecture features of Lingnan (south of Five Ridges in southern China); third, the exterior street module detail adopts Lingnan building materials, facade decorative lattice window and protruding canopy to enrich the modeling module; fourth, the detail modeling of European architecture is blended into the exterior street design as if the assimilation of Western culture into Lingnan culture; fifth, through combination and connection, different design modules are integrated into a whole, revealing the unique charisma of Lingnan culture at the same time; sixth, the landscape is themed with long sleeve dance of Cantonese opera, with flower beds, sculptures, fountains, etc. organically and undulantly laid out within the big curve texture.

04

WEIFANG WANDA PLAZA
潍坊万达广场

时间 2014 / 05 / 23 **地点** 山东 / 潍坊
占地面积 10.97 公顷 **建筑面积** 57 万平方米

OPENED ON 23rd MAY, 2014
LOCATION WEIFANG, SHANDONG PROVINCE
LAND AREA 10.97 HECTARES **FLOOR AREA** 570,000M²

OVERVIEW OF PLAZA
广场概述

潍坊万达广场项目位于潍坊市核心区（奎文区）老体育场位置。南北两侧分别为潍坊市最重要的两条东西向干道——东风东街和福寿东路，西侧为鸢飞路，东侧紧临虞河，地理位置优越。广场占地10.97公顷，总建筑面积57万平方米；项目由商业综合体、甲级写字楼、五星级酒店、商务酒店、住宅及商铺等业态构成。

Weifang Wanda Plaza is located at the original stadium's site in the core area (Kuiwen District) of Weifang City. With Dongfeng East Street and Fushou East Road, the two most important east-west arteries of the city, on the south and north sides, Yuanfei Road on the west and adjacent Yu River on the east, the Plaza enjoys an advantageous geographical location and covers a land occupancy of 10.97 hectares and gross floor area of 570,000 square meters. The project is composed of commercial complex, grade-A office building, 5-star hotel, business hotel, residence, stores and other formats.

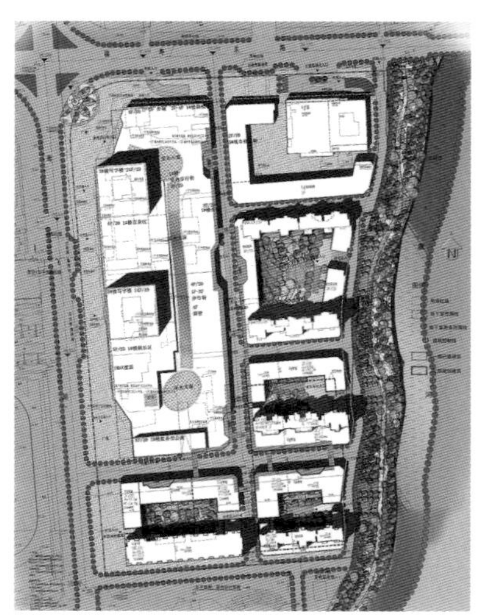

02

03

01 广场总平面图
02 广场全景
03 广场设计手绘稿

04

04 广场外立面
05 广场外立面设计手绘稿
06 广场外立面
07 广场百货外立面

05

FACADE OF PLAZA
广场外装

潍坊又称“鸢都”，风筝是潍坊最显著的特色。而一个以风筝为代表的城市，风成为这个城市最特别的元素。在设计中，极具力量感的折线形体块穿插组合，在深褐色铝板为背景的衬托下，斜指向天空，寓意吹动风筝扶摇直上九天的隐形力量——风。购物中心两侧主入口采用大面积的超白玻璃橱窗，简洁大方，强调了入口的可识别性。其中北侧主入口环绕富有层次感的彩釉玻璃，更加突出入口，使人们进入购物中心的感觉更加美妙。

Weifang is also known as the "Land of Kite", as kite is the most notable feature of this city. As for such a city represented by kite, wind, without any doubt, is the most special element of the city. Powerful broken line-shaped volumes, against the background of dark brown aluminum plates, point to the sky sideways, symbolizing wind, the invisible power that sends kites going flickeringly up the sky. Main entrance on both sides of the shopping center utilizes large ultra clear glass shop window, simple and elegant, increasing distinguishable characteristic of the entrance. And the main entrance on the north side is surrounded with layered enamelled glasses that further highlight the entrance and excite people when entering into the shopping center.

06

万达百货
WANDA DEPT STORE
刮刮卡领取
排队区

IMAX
BEANPOLE
ONLY
KEYROAD
2F
PEAC

08 圆中庭
09 商铺落位图
10 椭圆中庭

3F

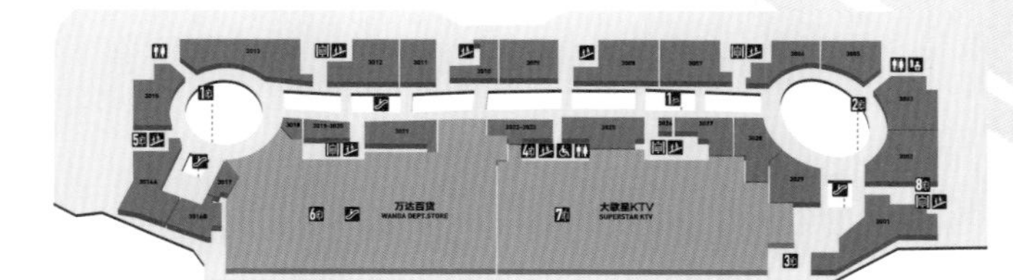

2F

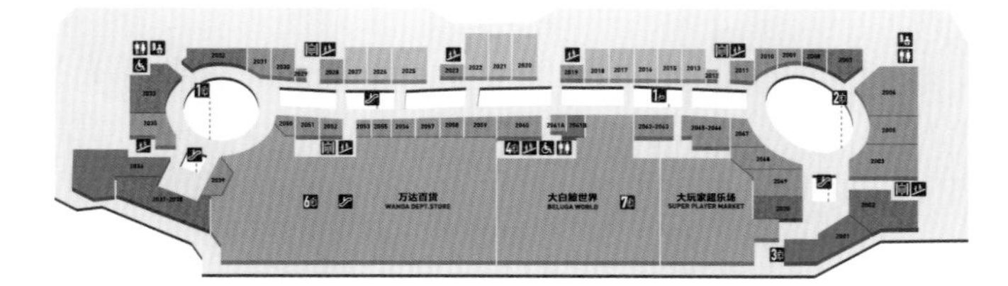

1F

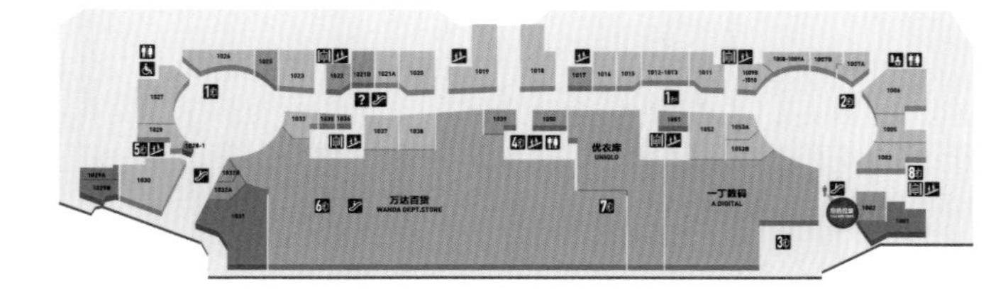

09

INTERIOR OF PLAZA
广场内装

内装在定位上把风筝的元素和潍坊人独有的生活方式有序地融入整个广场内部空间的设计中。室内的一些结构中充分体现了风筝的“放”和“飞”，也提炼了风筝的传统纹样作为装饰语言。

At the design orientation, both elements of kites and unique lifestyle of local people in Weifang city are also expected to be organically integrated into design of the whole interior space. Partial interior structures fully present “Fly” and “Flying” of a kite, and utilize refined traditional pattern of kite as decorative language.

10

LANDSCAPE OF PLAZA
广场景观

景观设计与建筑主题相呼应，在对风筝文化的提炼上摈弃了传统直白的处理手法，而重在体现飞翔轻盈的特性。设计运用折纸的形式来表现景观，营造出一种轻盈的氛围和飞翔的感觉，富有现代气息。“鸢飞九天”的主景雕塑仿佛是从地形中生长出来，火红色的腾飞感造型振翅翱翔，与水景相结合，并与建筑的流畅折线一起构成了潍坊的独特城市标志。

Landscape design of the Plaza, echoing the architectural theme, discards traditional straightforward treatment on refinement of kite culture and focuses on showing graceful flying of the kite. In the design, landscape is expressed in paper folding forms to build an atmosphere of lightness and feeling of flying, creating a modern design. The accent sculpture of "Flying Kite in the Sky" seems to grow up from the ground and the fiery red image fluttering and soaring high, together with the waterscape and flowing fold lines of building become the unique symbol of Weifang city.

11

12

13

NIGHTSCAPE OF PLAZA 广场夜景

11 主题雕塑
12 景观绿化
13 广场夜景
14 室外步行街

EXTERIOR PEDESTRIAN STREET 室外步行街

14

室外商业街立面取意潍坊市的象征风筝，色彩丰富，造型灵动，融合潍坊民居、后现代构成主义等多种建筑设计风格，力图打造极具潍坊地方特点的商业气氛。立面材料取材多样，石材、真石漆、清水砖、彩釉玻璃等搭配合理，错落有致。风筝的符号通过各种材料和形式的造型穿插其中，进一步烘托了设计主题，商业氛围和建筑立面设计达到高度统一。

Facade of the exterior commercial street, inspired by kite, symbol of Weifang City, has rich colors and vivid image and integrates multiple architectural design styles including dwellings in Weifang and post-modern constructivism in an attempt to build a commercial atmosphere full with local characteristics of Weifang city. Various kinds of facade materials, such as stone, stone-like coating, ganged brick and enamelled glass are reasonably combined and well-arranged. The symbol of kites is interspersed via various materials and types to further emphasize the design theme, achieving high unity between commercial atmosphere and facade design.

SHANGHAI SONGJIANG WANDA PLAZA
上海松江万达广场

时间 2014 / 05 / 30 **地点** 上海
占地面积 9.27 公顷 **建筑面积** 31.67 万平方米

OPENED ON 30th MAY, 2014 **LOCATION** SHANGHAI
LAND AREA 9.27 HECTARES **FLOOR AREA** 316,700M²

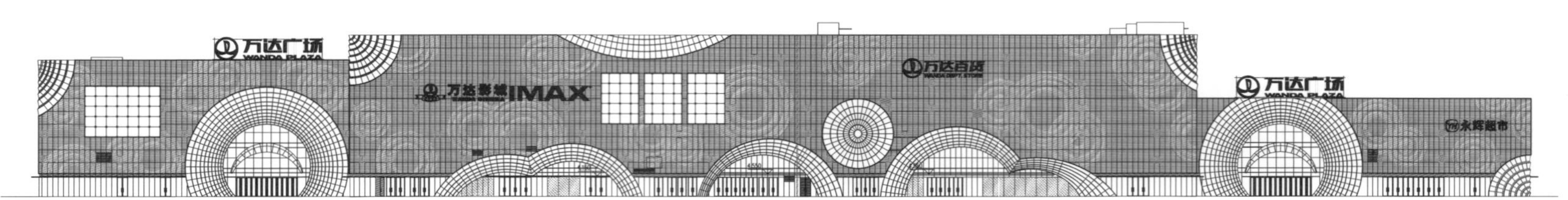

01

02

OVERVIEW OF PLAZA
广场概述

上海松江万达广场，位于“松江新城国际生态商务区”核心位置，由百货楼、娱乐楼和室内步行街、室外步行街及商务酒店、办公楼等组成，包括商业、文化娱乐、办公等业态。商业、文化娱乐部分高度为2~6层，4栋办公楼高度为19层，规划建筑高度不超过66米。酒店高度为10层，高度为41.05米，地下室为1层，另有局部的夹层为非机动车停车库。

Shanghai Songjiang Wanda Plaza is located at the core position of the "Songjiang New City International Ecological Business District", consists of department store, entertainment building, interior pedestrian street, exterior pedestrian street, business hotel, office building, etc., and involves commercial, entertainment and office activities. The commercial and entertainment area has two to six floors while the four office buildings 19 floors, with the planned building height being no higher than 66m. The hotel, with height of 41.05m, has ten floors including one floor of basement and partial mezzanine for non-motor vehicle parking.

01 广场立面图
02 广场全景
03 广场总平面图

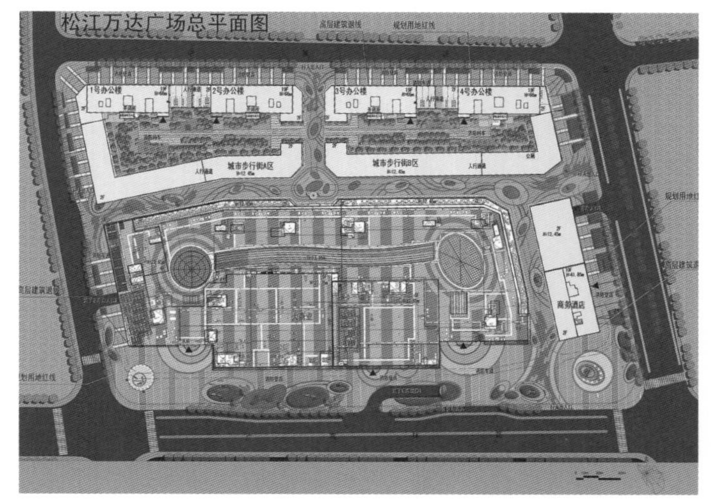

03

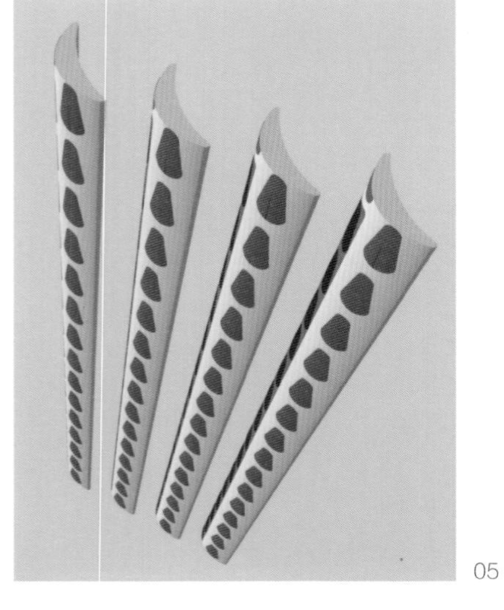

FACADE OF PLAZA
广场外装

松江历史悠久文化发达，是上海的文化之根，因水得名。整个大商业外立面恰如一个巨大的、平整的湖面，寓意松江之水，水滴成江。在室内商业街的主入口设计上，融合中国园林中“月亮门”的概念，利用玻璃层层叠退的手法使主入口形成旋涡一样视觉冲击力，吸引人流进入商场，成为强烈的视觉中心。

大商业主入口背衬百页，兼具实用与装饰作用，日间关闭遮挡强烈的日照，夜间打开展现室内的灯火通明。百页表面明暗组合的纹理，运用平面构图手法营造具有立体感的视觉效果，恰如月光倒影在微波的水面上，点缀主入口的月亮门，同时与大商业涟漪的主题隐隐地契合。大商业整个外立面利用穿孔铝板孔径的大小，形成一个个大小不一的水滴落湖的圈纹，与平静的湖面相结合，将“江南春雨”的特征完美地体现在建筑的外立面上。

04 主入口雨篷
05 主入口雨篷结构模块
06 主入口雨篷结构模块
07 广场外立面

07

Songjiang, with long history and developed culture, is the root of Shanghai culture and known for water. Whole facade of the large commercial area appears like a giant and smooth lake, meaning water of Songjiang converging river. Design of main entrance of the interior commercial street incorporates "Moon Gate" concept of classical Chinese garden and strengthens vortex like visual impact of the main entrance through step-back layers of glasses to attract people to enter in, forming an attractive visual center.

Main entrance of the large commercial area is lined with louvers, practical and ornamental, which will shut down during daytime as sunshade and open up as night falls to show the brightly lit interior space. Intertwined light and dark textures on the surface of louvers create a three-dimensional visual effect in the form of plane composition, just like moonlight glistening on the shimmering waters, decorating moon gate at the main entrance and echoing the ripple theme of large commercial area. Different pore diameters of perforated aluminum plates for facade of the whole large commercial area form lots of circles in different sizes on the facade like water dripping into lakes, and against the smooth

INTERIOR OF PLAZA
广场内装

广场内装以“精彩松江”为主题设计的理念，力求将松江万达广场室内步行街打造成时尚动感、富有当地文化情结的现代城市商业综合体形象。椭圆中庭侧板色彩的变化，星光点点，犹如晚霞般迷人，成为整个空间的视觉焦点，营造出动感绚丽、美轮美奂的空间氛围。

Following the theme design concept of “Wonderful Songjiang”, interior of the Plaza strives to build interior pedestrian street of the Shanghai Songjiang Wanda Plaza into a fashionable and dynamic modern urban commercial complex with love for local cultures. Side plates of the oval atrium, by changing the colors, like stars and the charming sunset glow, become visual focus of the whole space and create a gorgeous and incredible spatial atmosphere.

09

08

08 连桥
09 椭圆中庭
10 商铺落位图
11 直街

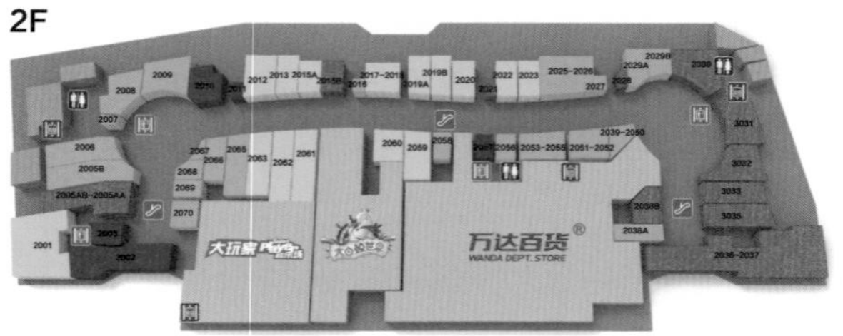

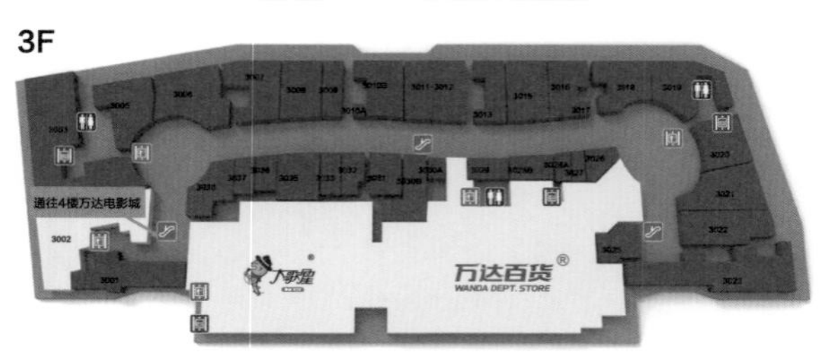

10

er Drink
避风塘

12

SUB-ANCHOR AND SPECIALTY CATERING
次主力店及特色餐饮

上海松江万达广场周边3平方公里内拥有密集的顾客资源——包括三湘、保利等7个成熟小区、昂立等商务楼及万科、信达、九龙仓等在建住宅项目；故上海松江万达广场商业布局设计了8家主力店、134家步行街商铺。独有特色的主力店吸客能力强，如大白鲸室内游艺城日均客流3000人次、万达影城IMAX厅日均客流3500人次，永辉超市日均客流15000人次。此外，餐饮楼层招商落位注重差异化的理念，拉宽了菜系的选择面，有效吸引了顾客。

Shanghai Songjiang Wanda Plaza has intensive customer resources within three square kilometers around the project, including Sanxiang, Poly, and some other five mature communities, Only business building, and some residential projects in progress, such as Vanke, Cinda, Wharf, etc. Given this, commercial layout of the Plaza is arranged with 8 anchor stores and 134 pedestrian street shops. Unique anchor stores have powerful customer attraction capacity, for example, daily customer flow of Beluga Indoor Entertainment, Wanda Cinema IMAX and Yonghui Supermarket respectively reaches 3,000 persons, 3,500 persons and 15,000 persons. Moreover, investment attraction and brand allocation of the catering floor emphasizes differentiation concept, thus broadening the range of cooking style and effectively attracting customers.

13

14

LANDSCAPE OF PLAZA
广场景观

东广场以《鹿跃云间》的主雕为中心，四周布置半围合的休憩座椅和立体花坛，形成一个可游可憩的商业广场空间。三只小鹿穿越圆环腾云而起，展示出奔腾活泼的动感，扭头的动作展示出“十鹿九回头”的故事场景。西广场的九色彩虹雕塑与建筑立面相呼应，两只小鹿悠闲地漫步其中，增添了广场的温馨气氛。这里不仅仅是时尚的购物广场，也是悠闲的生活广场。

The east plaza is spread around the theme sculpture of "Deer Leaping in Clouds" with semi-enclosed seats for rest and mosaiculture all around, forming a commercial square space where people can visit and rest. Three lovely deer jump through the ring and walk on clouds, showing visitors a lively dynamic picture while the action of looking back describing the scene of "Homecoming Deer". As for the west plaza, building facade is echoed by the nine-color rainbow sculpture with two strolling deer, further enhancing warm and sweet atmosphere of the plaza. Shanghai Songjiang Wanda Plaza is not only a fashionable shopping plaza but also a leisurely living plaza.

12 特色餐饮
13 次主力店店铺
14 次主力店店铺
15 广场景观

16 主题雕塑
17 景观绿化
18 广场夜景
19 室外步行街

16

17

NIGHTSCAPE OF PLAZA
广场夜景

购物中心立面穿孔铝板幕墙，简洁、时尚，通过利用孔径的大小来体现建筑圆环的肌理效果；灯光设计追随建筑肌理来布置灯光，在有圆环纹理的穿孔铝板内，固定安装LED点光源，旨在体现建筑纹理在夜间的形象；由于选择的灯体较小，半透明磨砂灯罩隐藏性好，对白天的立面效果几乎没有影响，夜间灯光跑动扩散，将“水滴涟漪”的自然景观效果，用灯光的方式在夜间融合到建筑夜景照明设计当中来，富有江南水乡的气息和意境。

Facade of the shopping center adopts perforated aluminum alloy curtain wall, simple and fashionable, on which holes in different sizes are distributed to show textural effect of the building circles. Moreover, lighting design is also arranged in accordance with the building texture, LED point light sources are fixed in perforated aluminum alloy plates with circular texture to show image of the building texture at night. Since selected lamps are small, translucent frosted lampshades can be well hided, causing little influence on the building facade effect in the day. As night falls, the spreading lights integrate the natural landscape effect of "Dripping Ripples" into nightscape lighting design of the building via lighting, presenting smells and sentiment of south of the Yangtze.

18

19

EXTERIOR PEDESTRAIN STREET
室外步行街

室外步行街在与大商业材料运用（铝板）统一的基础上，设计手法上有较大的反差，金街的整个立面设计整体很灵活，凹凸转折幅度较大，同时也在相同设计元素的基础上，利用不同的铝板的颜色，让整个的色彩感更加强烈，增加了商业街的商业效果，仿佛不同的多元文化融合在一起，形成绚丽多彩的整体效果。

Exterior Pedestrian Street follows material application of the large commercial area (aluminum plate) in a unified manner but breaks with distinctive contrast in design mode. Whole facade of the golden street appears to be flexible with relatively great turning and folding. Meanwhile, even for structures with the same design elements, aluminum plates in different colors are selected to strengthen the sense of color, enhance commercial effect of the commercial street and form a gorgeous and colorful overall effect with multi cultures integrated together.

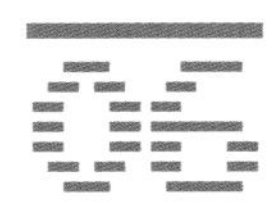

CHIFENG WANDA PLAZA
赤峰万达广场

时间 2014 / 06 / 20 **地点** 内蒙古 / 赤峰
占地面积 27.14 公顷 **建筑面积** 98.32 万平方米

OPENED ON 20th JUNE, 2014
LOCATION CHIFENG, INNER MONGOLIA AUTONOMOUS REGION
LAND AREA 27.14 HECTARES **FLOOR AREA** 983,200M^2

01 广场外立面
02 广场总平面图

01

OVERVIEW OF PLAZA 广场概述

02

赤峰万达广场项目位于赤峰市新城区，北至锡伯河南岸，南至西拉沐沦大街，东至平双公路，西至宝山路所围合的范围。总规划用地面积27.14公顷，其中A区为五星级酒店、甲级写字楼、三星级酒店、商业街、住宅；B区为购物中心、商业街、住宅；C区为写字楼、商业街住宅。

Chifeng Wanda Plaza is located in the new town of Chifeng, north to south bank of Xibe River, south to Xilamulun Street, east to Pingshuang Road (306 National Road) and west to Baoshan Road. The Plaza has a total planning land area of 27.14 hectares, including three zones, respectively 5-star hotel, grade-A office building, 3-star hotel, commercial street and residence for Zone A, shopping center, commercial street and residence for Zone B and office building, commercial street and residence for Zone C.

03

FACADE OF PLAZA
广场外装

赤峰万达广场的外立面设计将内蒙古“草原”这一地域特点贯穿始终，无论从建筑形式的推敲还是表皮肌理的处理，无处不体现“草原”这一灵动的自然元素。表皮肌理运用三维立体白色铝板，以流线为基本元素，勾勒出“草原”的脉络。整个建筑将草源河川生机勃勃的景象演绎得淋漓尽致。门头造型创意取自一条蜿蜒曲折的江水似银龙般流过无边无际的绿色草原。逐层渐退的门框形式有着丰富的层次感，玻璃与铝板间的虚实的变换、视觉元素的重现带来强烈的视觉冲击。

Facade design of Chifeng Wanda Plaza consistently follows regional characteristic of “Grassland” of Inner Mongolia, and such vivid natural element is fully reflected whether from careful consideration of architectural form or processing of surface texture. Surface texture of the Plaza takes streamline as the basic element and utilizes 3D white aluminum plates to sketch veins of the “grassland”. The building incisively and vividly presents visitors the active and vibrant grassland and streams. Shape design of the gate is evolved from the image of one wandering river, just like a silver dragon, flowing through the boundless green grassland. The layer-by-layer step-back door frames create differences in gradation, while changing between false & real of glass & aluminum plate and reappearing of visual elements generate strong visual shock.

04

05

06

07

08

03 广场外立面
04 广场百货外立面
05 广场主入口
06 广场外立面特写
07 广场外立面特写
08 广场外立面特写

09

10

INTERIOR OF PLAZA
广场内装

圆中庭采用了体块穿插的手法，简约的模数化排板、体块造型贯穿整个中庭，弧线体块在空间中穿插，使得圆中庭整体感观浑然一体。圆中庭侧板分为两种材质形态——透光肌理的人造石与GRG材质，摈弃了繁琐的装饰元素，简约而时尚。椭圆中庭，围绕观光电梯周围的侧裙板采用整体衔接表现手法，利用一到三层大体量的造型关系，增强了中庭强大的视觉效果；侧板透光效果的细节处理，以及漂浮设计手法，营造出轻盈灵动的空间特色。

The circular atrium adopts the cross technique of blocks. The simple modular plates and blocks running throughout the whole atrium and curved blocks interspersed in the space endow the circular atrium with an overall feeling as an integrated whole. The side plates used in the circular atrium have two types of materials and shapes, specifically artificial stones with translucent texture and advanced GRG materials, and rejects tedious decorative elements, appearing to be simple and stylish. The oval atrium, centering on side plates around the sightseeing elevator, adopts expression way of integral connection and takes advantage of the large volume shape spreading from the ground floor to the second floor to strengthen shocking visual effect of the atrium. Moreover, details like light transmission of the side plates and floating design skill create a delicate and flexible space feature.

12

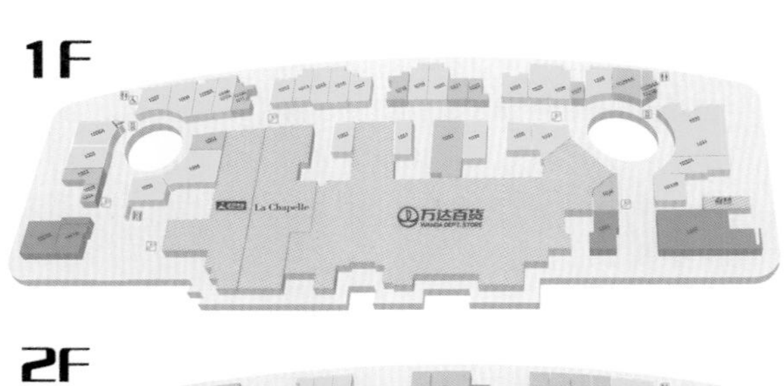

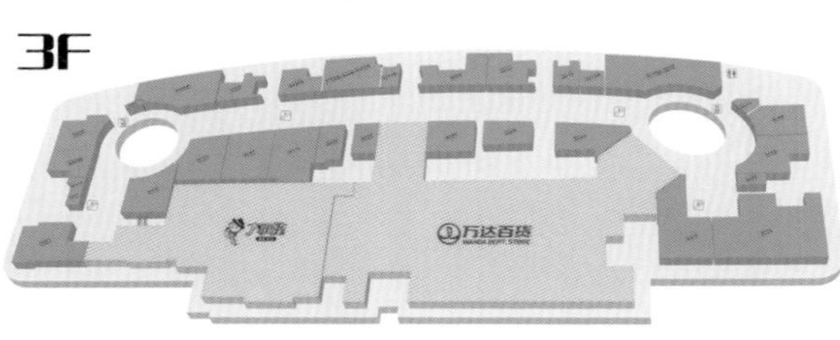

11

服装 精品 餐饮 体验

09 圆中庭天花
10 圆中庭
11 商铺落位图
12 室内步行街

13 主题雕塑
14 广场地铺
15 广场夜景
16 室外步行街

LANDSCAPE OF PLAZA
广场景观

景观方案的设计理念来自赤峰红山，铺装的肌理提取自红山冰川时期的岩石断层质感，暖色系的铺装凸显出赤峰红山的红，铺装的带状元素提取自建筑的竖向线条。主入口广场的梯形铺装象征着赤峰主峰，立体化的花坛小品的设计吸取了建筑上的弧形元素结合人体工程学设计。雕塑取材赤峰当地的著名玉龙文化，结合现代的雕塑工艺打造出具有现代感的玉龙文化雕塑。

Design concept of the landscape scheme is evolved from the Hongshan of Chifeng, texture of the pavement comes from fault texture of rocks of the Hongshan in the glacial period and band elements of the pavement are extracted from vertical lines of the building. Warm-colored pavement highlights red color of the mountain and trapezoidal pavement at main entrance of the Plaza symbolizes the highest peak of Chifeng. The mosaiculture design assimilates arc elements in architectural design together with ergonomic design, and sculptures, originated from local famous jade dragon culture in Chifeng, combines modern technology to craft sculptures, showing jade dragon cultures in modern style.

NIGHTSCAPE OF PLAZA
广场夜景

15

EXTERIOR COMMERCIAL STREET
室外商业街

16

赤峰万达广场结合独特的用地现状，购物中心仅在东西两侧设置自持商铺，无室外金街及一环铺。因此，室外自持商铺外立面需要保持购物中心立面风格的整体性的同时，需要体现独立商铺的特征。商铺一层延续购物中心橱窗的通透性，采用通透的玻璃展示立面；二层增加特色店招位置，凸显商业氛围及店铺特征。

In view of distinctive land conditions of Chifeng Wanda Plaza, the shopping center only arranges holding stores on the east and west sides without golden street or first ring stores. Therefore, facade of the exterior holding stores should maintain unified style of the shopping center and prominent characteristics of the independent stores as well. The ground floor, following transparency of shop windows of the shopping center, adopts transparent glass showing facade; while the first floor is additionally arranged with locations for feature signage to highlight the commercial atmosphere and store features.

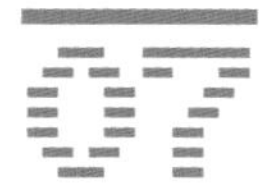

JINING TAIBAI ROAD WANDA PLAZA
济宁太白路万达广场

时间 2014 / 07 / 05 **地点** 山东 / 济宁
占地面积 12.98 公顷 **建筑面积** 73.67 万平方米

OPENED ON 5th JULY, 2014
LOCATION JINING, SHANDONG PROVINCE
LAND AREA 12.98 HECTARES **FLOOR AREA** 736,700M²

01 广场外立面
02 广场立面图
03 广场外总平面图

01

02

OVERVIEW OF PLAZA
广场概述

济宁太白路万达广场位于济宁市太东工贸有限公司以东、刘庄路以南、琵琶山路以西、太白东路以北，占地12.98公顷，总建筑面积73.67万平方米，地上建筑面积59.65万平方米，地下14.02万平方米。广场集大型商业购物中心、室外商业街、精装SOHO、五星级酒店、甲级写字楼、高档住宅为一体，成为引领鲁南商业潮流的现代化城市综合体。商业部分总面积20.27万平方米，购物中心地上8.11万平方米，地下7.6万平方米，商铺4.56万平方米。

Located to the east of Jining Taidong Industrial and Trading Co., Ltd, south of Liuzhuang Road, west of Pipashan Road and north of Taibai East Road, Jining Taibai Road Wanda Plaza covers a land area of 12.98 hectares and overall floor area of 736,700 square meters, including aboveground floor area of 596,500 square meters and underground floor area of 140,200 square meters, and integrates large-scale shopping center, exterior pedestrian street, refined-decorated SOHO, 5-star hotel, grade-A office building and high grade residence into one, becoming a modern urban complex leading the commercial trend of south Shandong. Commercial part of the Plaza occupies a total area of 202,700 square meters, the shopping center covers 81,100 square meters for the aboveground part and 76,000 square meters for the underground and the store area is 45,600 square meters.

03

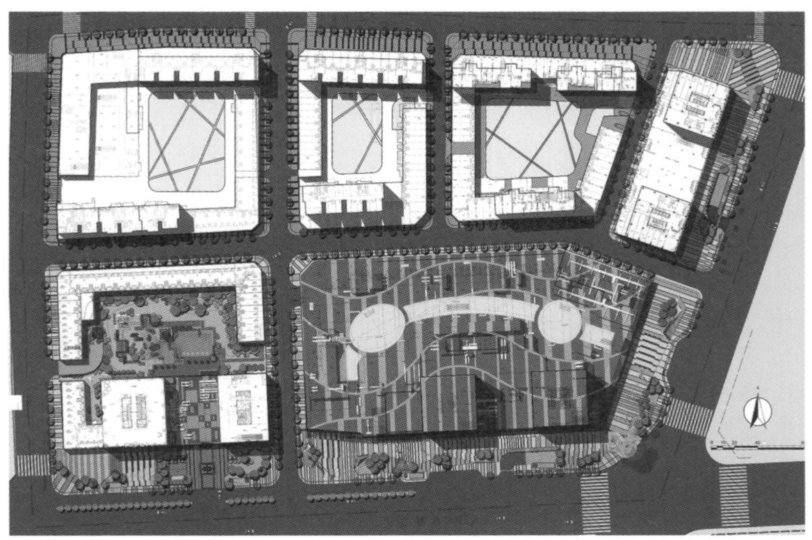

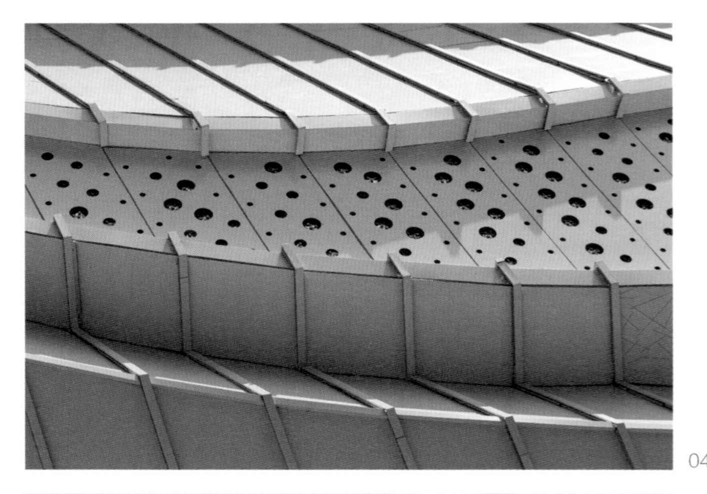
04

05

FACADE OF PLAZA
广场外装

商业立面造型立意取自京杭大运汹涌澎湃的河流，使整个商业立面气势恢宏又具有现代气息。圆点状的穿孔板穿插在波纹状的墙面上，增强整体动感，活跃商业氛围；抽象的波浪形成巨大拱形，彰显商场入口的趣味性、识别性和引导性。

Originated from surging Grand Canal, facade of the whole commercial part appears to be modern but magnificent. Dot-shaped perforated plates intertwined on corrugated walls enhance the integral dynamic and create active and lively commercial atmosphere, and abstract waves-formed huge arch manifests interestingness, identity and guidance of the department store entrance.

04 广场外立面局部
05 广场外立面局部
06 广场外立面

06

INTERIOR OF PLAZA
广场内装

广场室内设计从济宁市市花“荷花”及运河中得到设计母题，以极具时代感的金属、玻璃、石材等材质把荷花、运河水的元素通过提炼、简化和抽象处理的图案展开设计。通透、简洁，洋气的中庭波浪弧面，造型独特的观光电梯是圆形中庭的焦点；三层的中庭侧裙以写意的手笔，轻轻带出运河之水的造型；再辅以香槟金色的材质搭配，体现出灵动、高雅之感。以市花“荷花”为椭圆中庭的设计主题，地面石材拼花采用流畅的花瓣造型，非常优美；侧裙通过穿孔发光铝板图案的手法把花瓣造型隐隐约约地勾勒出来，光影效果变化丰富，整个中庭非常的时尚、高雅。

Interior design of the Plaza, with theme derived from lotus, the city flower of Jining and canal, utilizes metal, glass and stones with strong sense of times to refine, simplify and abstract the elements of lotus and canal water into patterns for design. The translucent, simple and fashionable waved cambered surface of atrium and unique sightseeing elevator are visual focus of the circular atrium. Side skirt of the atrium on the second floor traces out image of the canal water in freehand sketching and champaign gold materials helps to create a dynamic and elegant sense. Based on the design theme of "Lotus", floor stone pattern of the circular atrium adopts flowing petal as the model, showing visitors a graceful view, and side skirt sketches out the petal shape through perforated lighting aluminum plates to form a changing effect of light and shade, building a fashionable and elegant atrium.

1F

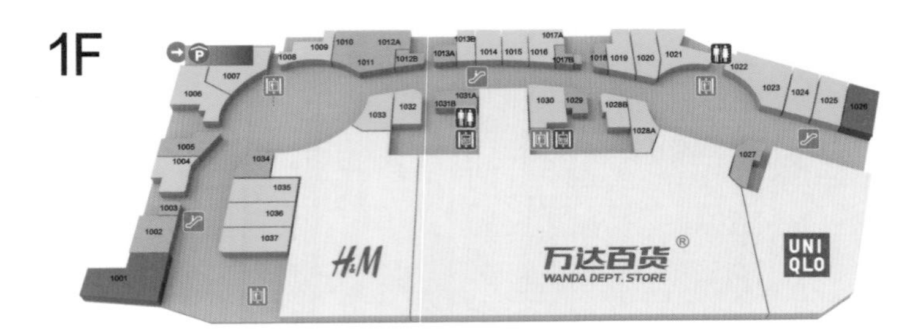

2F

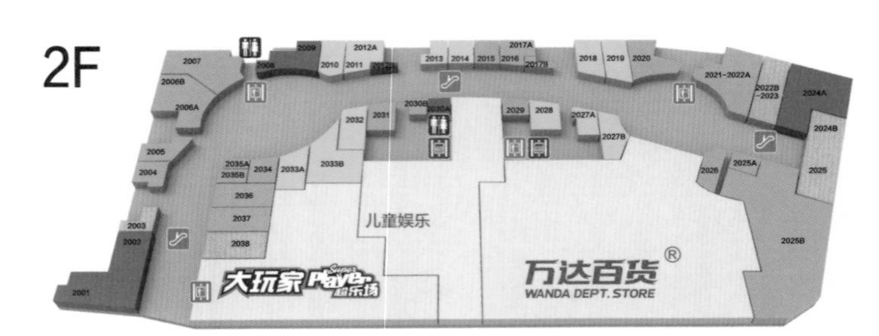

3F

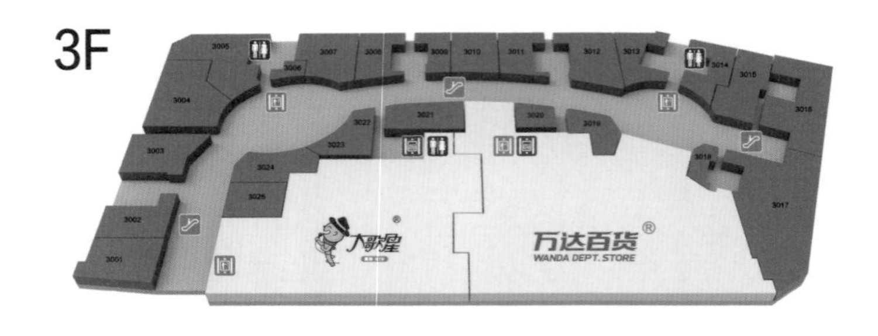

服装服饰 生活精品 体验式服务 餐饮美食

09

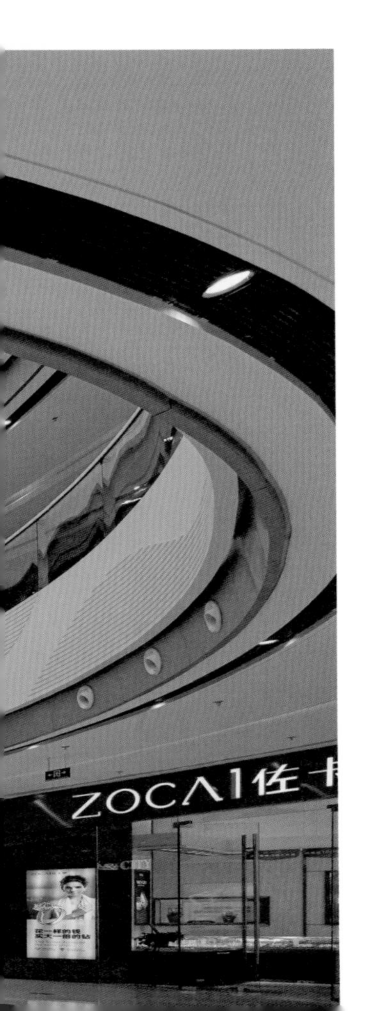

07 商铺落位图
08 椭圆中庭
09 椭圆中庭
10 室内步行街滚梯

10

11

12

13

14

LANDSCAPE OF PLAZA
广场景观

景观设计提取了济宁最有代表性的两个文化元素：儒家文化与运河文化。将竹简、书法和波浪等符号与建筑相结合，凸显万达广场在当地“运河上流淌的悠久文明，都市中跃动的明日之星”的地位。

Landscape design of the Plaza is conceived from Confucian culture and canal culture, the two most representative cultural elements of Jining. Bamboo slips, calligraphy, wave and similar symbols are combined with architecture to reflect the status of Wanda Plaza as "Time-honored Civilization Flowing on the Canal and Rising Star Beating in the City" in Jining.

11 景观小品
12 景观小品
13 景观小品
14 景观小品
15 广场夜景
16 室外步行街

15

EXTERIOR PEDESTRIAN STREET
室外步行街

采用的是弧形的元素组合，完全与大商业相呼应，为丰富金街的商业氛围，在色彩设计、组合上更加注重其活泼性。白色与灰色的渐变、蓝色渐变及橙色与黄色的组合，均为丰富金街的手法。

Exterior Pedestrian Street adopts combination of arc-shaped elements echoing with the large commercial area and emphasizes liveliness and vividness in color design and combination to enrich commercial atmosphere of the golden street, such as color gradient of white & gray and blue and color combination of orange and yellow.

16

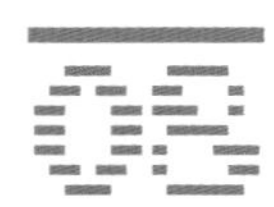

JINHUA WANDA PLAZA
金华万达广场

时间 2014 / 07 / 25 **地点** 浙江 / 金华
占地面积 9.08 公顷 **建筑面积** 50.59 万平方米

OPENED ON 25th JULY, 2014
LOCATION JINHUA, ZHEJIANG PROVINCE
LAND AREA 9.08 HECTARES **FLOOR AREA** 505,900M²

01 广场鸟瞰
02 广场外立面
03 广场总平面图

01

02

OVERVIEW OF PLAZA
广场概述

金华万达广场位于浙江省金华市李渔大桥东，总建筑面积50.59万平方米，其中地上40.54万平方米，地下10.05万平方米，涵盖了高端购物中心、城市室外步行街、五星级酒店、甲级写字楼和高级公寓等业态。

Jinhua Wanda Plaza is located to the east of Liyu Bridge in Jinhua, Zhejiang Province, covers an overall floor area of 505,900 square meters, including 405,400 square meters for the aboveground part and 100,500 square meters for the underground and integrates high-end shopping center, exterior urban pedestrian street, 5-star hotel, grade-A office building and high-end apartment, etc.

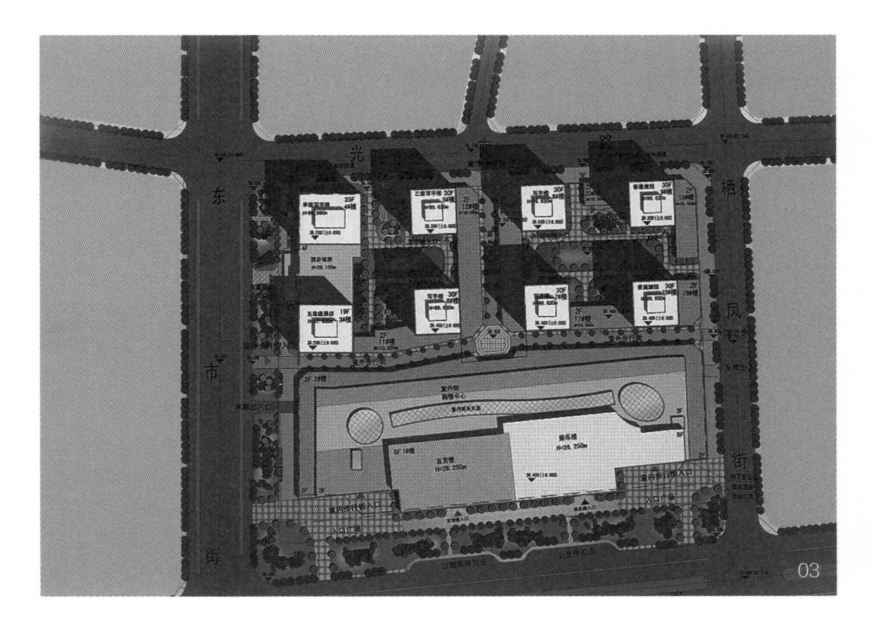

03

04

05

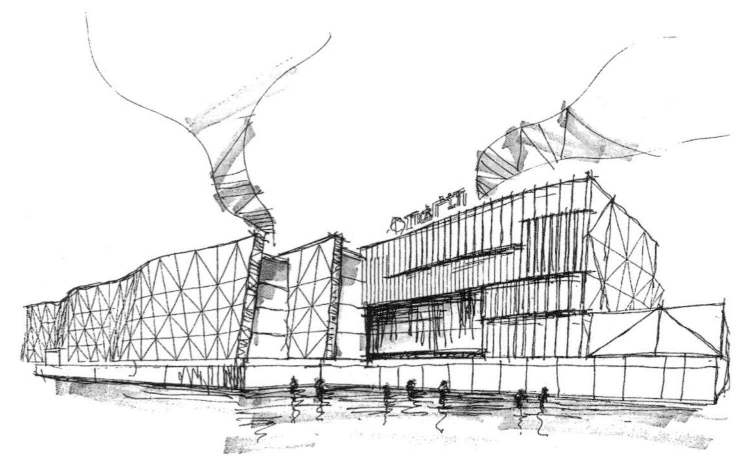
06

04 广场外立面
05 广场局部立面
06 广场外立面设计手绘

FACADE OF PLAZA
广场外装

金华婺江由义乌江和武义江在金华城南汇流为一，沿岸风光亮丽，气势雄伟，带给了金华整个城市灵气，增添了金华江南秀丽的城市气质。广场在形体上取意于双溪柔和蜿蜒的造型，前后两条曲板，按相同曲率扭动生成。曲线的造型丰富了建筑形体的天际轮廓线，宛若一湾流动的溪水，暗喻了双溪柔美曲折的形态特色。入口采用三角折板按照不同曲线阵列所成的肌理，通过变化三角的角度，折板也会相应产生不同方向的韵律——在阳光的照耀下，随反射角的不同，折板也会随之变化呈现不同的颜色，从而产生丰富多变的立面形态。

Benefiting from beautiful scenery along the coast and its imposing manner, the Wujiang River, confluence of Yiwu River and Wuyi River at Chengnan, Jinhua, adds vitality to the whole city and enhances beauty of Jinhua as a city in South China. The Plaza's shape design, conceived from the shape of gentle and winding Shuangxi (currently Yanweizhou), is formed by two curved panels in the front and at the back twisted in the same curvature. Such curve shape enriches flowing stream-like skyline of the building and represents features of the winding Shuangxi. Entrance of the Plaza adopts triangle folded plates arrayed in different curves to build the texture. Through adjusting the angles, the folded plates will generate rhythm in different directions, that is, plates under the sunshine will present different colors along with different angles of reflection and consequently create an ever-changing facade.

07

08

INTERIOR OF PLAZA
广场内装

广场内装的主题及设计元素均以金华佛手为依据，通过造型元素的综合运用，形成简练而丰富的空间内容。圆中庭以金色年华为大基调，侧帮采用金色冲孔铝板辅以白色石膏板完美融合了刚性与柔性的对接，金色铝板的排序方式采用构成学中的渐变方式，符合了当代设计审美观念。金色冲孔铝板内设有LED灯，夜晚，从板孔放射的灯光散发出独特的韵味。地面石材运用流线造型进行无缝拼接，释放出时间的概念，也寓意出对金色年华的设计理解。椭圆中庭侧帮造型弧形更加圆滑与自然，毫无突兀感。弧形造型提取了从金沙遗址出土的太阳神鸟内层造型元素，赋予了金华万达广场特殊的意义。

Theme and elements of interior design of the Plaza are determined based on Jinhua fingered citron. Comprehensive application of various modeling elements helps to form concise but rich spatial content. The circular atrium, keeping to the keynote of Golden Times, adopts golden punched aluminum plates and auxiliary white plaster board for the lateral wall to reach seamless jointing of rigidity and flexibility. The golden punched aluminum plates are arranged in a gradual changing way that coincides with modern design aesthetic concepts. As night falls, LED lamps mounted in the golden punched aluminum plates will send forth unique and special flavor. Floor stones are seamlessly paved in a streamline style, stating conception of time and also expressing understandings on the design of golden times. As for the oval atrium, arc curved design of the later wall appears to be more smooth and natural. This shape design is inspired by inner modeling elements of the Golden Sun Bird excavated at Jinsha Site and endows the Jinhua Wanda Plaza with special significance.

07 圆中庭
08 室内步行街
09 商铺落位图
10 室内步行街

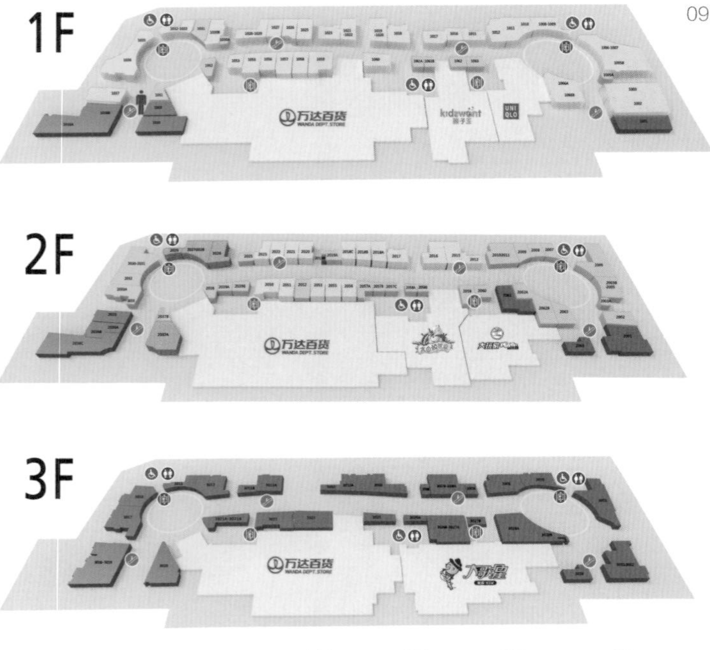

09

Lee

LANDSCAPE OF PLAZA
广场景观

景观设计以金色年华为概念设计主题，将建筑的表皮元素引申为金色带曲面的一张船帆，与地面景观相互融合成就一段金色航程，同时引入各种形式的海洋和水的元素。文化主题植入以“金华传统剪纸”为切入点，将剪纸的元素引入商业景观中，增添了地域文化氛围。

Centering on the conceptual design theme of Golden Times, landscape design of the Plaza extends surface elements of the building to a golden sail with curved surface to start a golden voyage integrated with the ground landscape. Meanwhile, elements of ocean and water in all kinds of types are introduced in the landscape design. Cultural theme placement takes Jinhua traditional paper-cutting as the breakthrough point, through which element of paper cutting is integrated with the business landscape, enhancing regional culture atmosphere.

NIGHTSCAPE OF PLAZA
广场夜景

13

EXTERIOR PEDESTRIAN STREET
室外步行街

14

15

金华万达室外步行街立面处理以连续性以及节奏感为设计原则，模块化是室外步行街的一大特点。这种拼接方式既丰富了外立面的形式，也降低了施工难度。各单元体量根据位置的不同进行颇有趣味的前后、虚实、穿插处理，构图均衡，在建筑表皮上形成不同的立面深度，营造出"一店一色"的商业效果；在加强商业氛围和强调重点空间的同时也形成了宜人的尺度。石质方框、木质排架、密排格栅、菱形小窗、米字型图案玻璃，似乎都在诉说着与金华这座城市千丝万缕的联系。

Facade of the exterior pedestrian street is treated based on the design principle of continuity and rhythm and modularization is one major characteristic of the exterior pedestrian street. This kind of splicing mode not only enriches forms of the facade, and also reduces construction difficulties. Each unit volume, according to different locations, undergoes interesting front and back, virtual and real and cross processing and achieves balanced composition to form various facade depths on the building surface, creating the commercial effect of "one shop, one feature" and builds pleasant scale while strengthening commercial atmosphere and highlighting important spaces. All elements, such as the stone square frames, timber frame bents, close packed grilles, diamond-shaped windows and star-shaped glasses, seem to recount their innumerable relationships with the city, Jinhua.

CHANGZHOU WUJIN WANDA PLAZA
常州武进万达广场

时间 2014 / 08 / 08 **地点** 江苏 / 常州
占地面积 12.0 公顷 **建筑面积** 60 万平方米

OPENED ON 8th AUGUST, 2014
LOCATION CHANGZHOU, JIANGSU PROVINCE
LAND AREA 12.0 HECTARES **FLOOR AREA** 600,000M²

OVERVIEW OF PLAZA
广场概述

常州武进万达广场位于常州市武进区湖塘花园街，占地12.0公顷，总建筑面积60万平方米，其中地上47万平方米，地下13万平方米，是一个包含大型购物中心、室外时尚步行街、五星级酒店、甲级写字楼、高级公寓及高档住宅等多功能于一体的城市综合体。

Changzhou Wujin Wanda Plaza is located on the Huayuan Street, Hutang Town, Wujin District, Changzhou, and covers a land area of 12.0 hectares and an overall floor area of 600,000 square meters, including 470,000 square meters for the aboveground part and 130,000 square meters for the underground. The Plaza is an urban complex that integrates large shopping center, exterior fashionable pedestrian street, 5-star hotel, grade-A office building and high-end apartment and residence, etc. into one.

02

03

01 广场总平面图
02 广场全景
03 广场设计手绘图

永辉超市
Bravo
YH
JUCYJUDY
hotwind
04

05

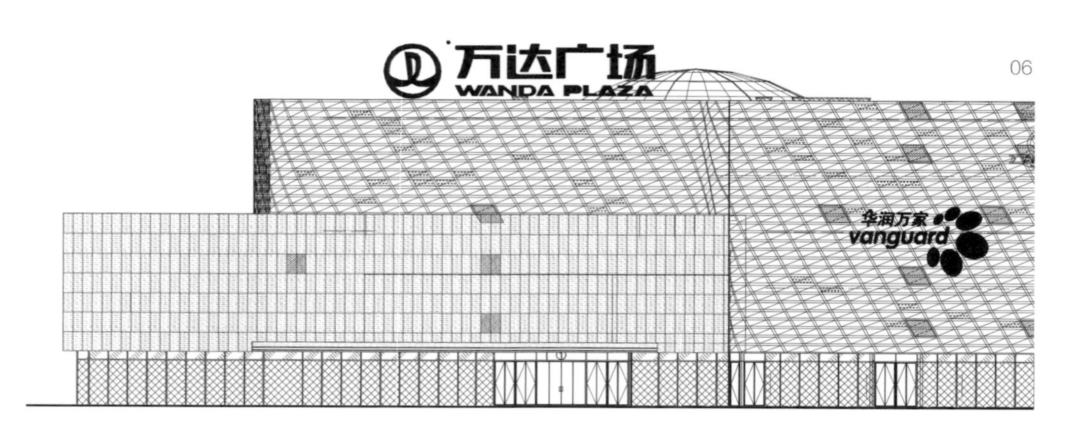
万达广场
WANDA PLAZA
华润万家
vanguard
06

FACADE OF PLAZA
广场外装

常州市位于江苏省南部美丽富饶的长江三角洲中心地带，别名“龙城”。武进是常州市的一个重要组成部分，“吴文化”的发源地之一。建筑设计以“龙”及地方文化为切入点，以“龙鳞”为基本单位，龙形为构图元素，以地方文化为基本的色调，从而充分体现建筑的地域性。建筑入口设计结合大商业整体造型，通过“减法”形成建筑入口，简洁大气，将主入口两侧适当高度内的“玻璃-金银折板”双层幕墙以二维曲线向内侧弯曲，衔接主入口处与首层玻璃幕墙齐平的直面，内收上方的部分则保持流畅的曲线造型，从而形成富有进入感的主入口。

Changzhou, also known as the "Dragon Town", lies in central zone of the beautiful and wealthy Yangtze River delta, at the south of Jiangsu Province. As an important part of Changzhou, Wujin is one of the places where "Wu Culture" originated. Architectural design of the Plaza takes "Dragon" and local culture as the breakthrough point, "Dragon Scale" as the fundamental unit, dragon shape as the compositional element and local culture as basic tone to fully reflect regional characteristics of the building. The concise and grand entrance design against the background of overall shape of the large commercial area builds a simple but grand building entrance in "subtraction" mode. Double-layer curtain wall made of "glass-gold and silver folded plate" within appropriate height on both sides of the main entrance are bent inwardly in two-dimensional curve to connect with plane of the main entrance level with the ground floor glass curtain wall, while upper part of the inwardly bent section still keeps the smooth curve shape in an attempt to build an attractive main entrance.

04 广场立面近景
05 广场立面远景
06 广场立面图

07

INTERIOR OF PLAZA
广场内装

圆中庭采用民族图案的打散重组，形成现代感极强的侧裙装饰元素，配合大的侧裙走线形成交叉错落的立体视觉对位，形成空间的韵律美；大面积的白色更加能够烘托商业店铺的特色和精彩。天花的造型是元素的简化，点缀有点单调的天花板，给人以近尺度的观赏，整体风格鲜明突出，容易给人留下难忘印象，成为点缀、衬托商业的绿叶。

The circular atrium takes advantage of separation and reorganization of national patterns to create modern side skirt decorative elements, forming staggered stereo visual alignment together with the obvious side skirt line and endowing the space with a kind of rhythm beauty. The widespread white can further emphasize features and characteristics of stores. Shape of the suspended ceiling is simplification of elements and embellishes the singsong ceiling to offer visitors nearby view. Overall style of the suspended ceiling is quite distinctive and will easily impress visitors, becoming embellishment and foil of the commercial environment.

08

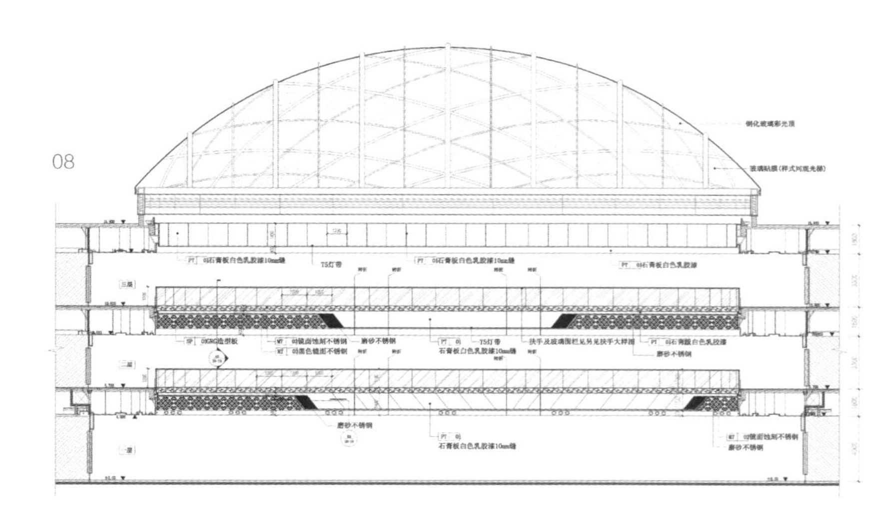

1

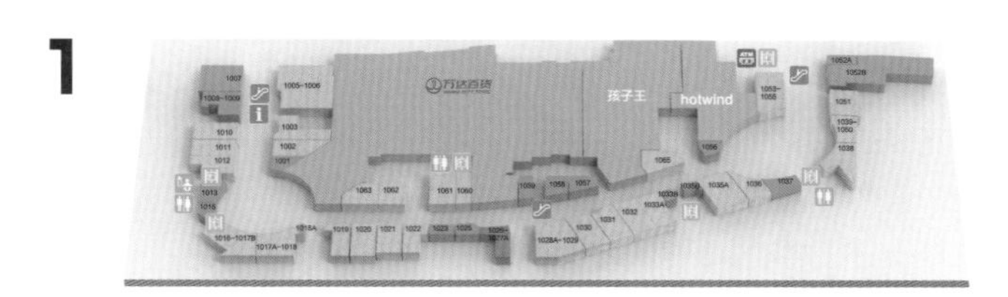

2

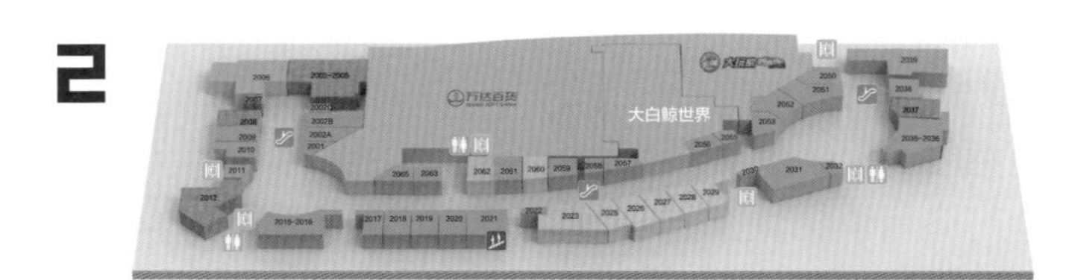

3

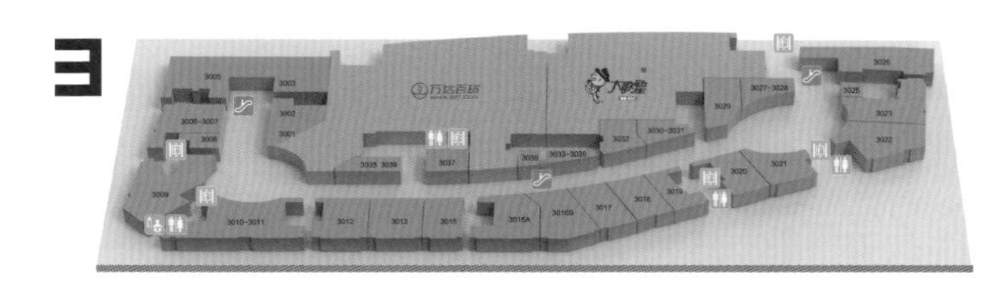

09 服装服饰 生活精品 体验式服务 餐饮美食

10

07 圆中庭
08 椭圆中庭剖面图
09 商铺落位图
10 室内步行街

11

LANDSCAPE OF PLAZA
广场景观

主广场人流相对集中，为形成较为开阔的集散区域，在大尺度的开放空间下，靠边点缀种植高大乔木，行成视觉的引导，加强纵向空间感的同时又能衬托主体建筑。高大乔木搭配自然种植、丰富多层次的下层灌木地被，形成广场前一道靓丽的绿化景观，并把灌木的高度控制在1.5m以下，以保持人视线的通透性，不对后方的建筑造成遮挡。在紧贴建筑侧墙面的绿化位置，通过浓密多层次的植物搭配进行遮挡，有效地软化生硬的墙体。

The main square has relatively dense stream of people, therefore, in order to build an open distributing space, tall and straight trees are planted to the side in large-scale open spaces to be served as visual guidance. This arrangement will enhance the longitudinal sense of space and simultaneously set the major building off. The tall and straight trees together with natural and diverse shrub ground cover build a beautiful greenery landscape in front of the square. Meanwhile, the shrubs are controlled within the height of 1.5m to guarantee an unobstructed view and avoid shading buildings behind. Dense and multi-leveled plants are planted at the greenery points closely adjacent to side wall of the building for shading, thus effectively softening the rigid wall.

12

NIGHTSCAPE OF PLAZA 广场夜景

13

11 主题雕塑
12 广场景观
13 广场夜景
14 室外步行街

EXTERIOR PEDESTRIAN STREET 室外步行街

从“一店一色、自然生长”的商业街设计概念出发，以现代立面为基础，用丰富材料、形式和颜色，将典型单元式模块和特殊的单元式模块有机组合，去营造一个刚开业就有浓郁历史感和情调的商业步行街，实现了“完全模拟原生态自发形成商业立面形态”的建设构思。建筑立面利用混搭的建筑形式、材料的多元化对建筑细部刻画，以及雨篷、壁灯、店招等元素的整体设计，形成了室外步行街独特的建筑风格。

Proceeding from the design concept of "One Shop One Style, Natural Developing" and based on modern facade, the exterior pedestrian street design organically combines typical unit modules and special unit modules together by abundant materials, forms and colors to build a commercial pedestrian street that is full of historical sense and sentiment since its opening, realizing the building concept to "completely simulate original ecology and spontaneously form commercial facade". The building facade utilizes combined architectural forms and diversified materials for details processing, plus the overall design of canopy, wall lamp and signage, endowing the exterior pedestrian street with unique architectural style.

14

10

FOSHAN NANHAI WANDA PLAZA
佛山南海万达广场

时间 2014 / 08 / 29 **地点** 广东 / 佛山
占地面积 9.7 公顷 **建筑面积** 70 万平方米

OPENED ON 29th AUGUST, 2014
LOCATION FOSHAN, GUANGDONG PROVINCE
LAND AREA 9.7 HECTARES **FLOOR AREA** 700,000M²

OVERVIEW OF PLAZA
广场概述

佛山南海万达广场位于佛山市南海区广东金融高新技术服务区的核心区段，用地北面紧贴广佛地铁金融城站。广场占地9.7公顷，总建筑面积70万平方米（地上逾53万平方米，地下逾16万平方米）。建筑群体分为南北两个区域——北区为住宅区，南区为公建区。公建区包括购物中心、产权式酒店、室外商业街，1栋高达185米44层超高层甲级写字楼和1栋29层甲级写字楼。

Foshan Nanhai Wanda Plaza is located at core section of Guangdong High-Tech Service Zone for Financial Institutions in Nanhai District of Foshan City, close to Financial City Station of the Guangfo Line on the north. The Plaza covers a land area of 9.7 hectares and overall floor area of 700,000 square meters including more than 530,000 square meters for the aboveground part and over 160,000 square meters for the underground part. The building complex can be divided into south and north zones, respectively the residential area for the south zone and public building area for the north zone, of which the public building zone is composed of shopping center, property hotel, exterior commercial street, one 185m tall and 44-floor super high-rise grade -A office building and one 29-floor grade-A office building.

01

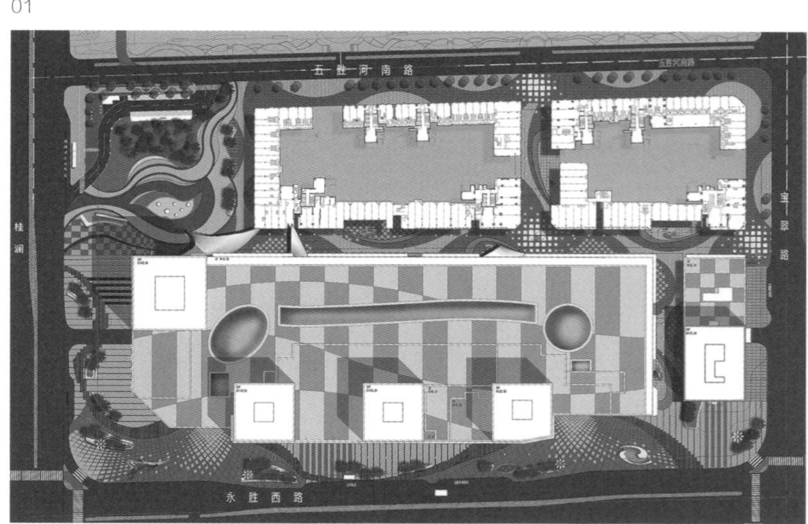

01 广场总平面图
02 广场外立面

04

FACADE OF PLAZA
广场外装

在大商业主体立面设计上，表皮意向采用横竖条纹的变化，材质简洁纯粹，但在光影下却产生出丰富的明暗变化肌理效果。建筑群立面设计手法统一，结合了飘带与花的设计意向，通过在空间中的错动、扭转的线条形成优雅的双曲面表皮变化。金属表皮的横竖向条纹肌理变化，结合飘带大压花玻璃质感的选择，在不同角度光影下会产生丰富的外立面效果，从而进一步增强了购物中心的商业气氛。

As for facade design of main structure of the large commercial area, the surface intends to adopt varied cross and vertical lines. The materials are simple and pure, but are able to produce textural effect with rich and varied bright and dark changing. Facade design of the building complex is completed in a unified manner and combines design intention of ribbon and flower, creating an ever-changing elegant double curved surface through moving and twisting lines in the space. Textural changes of varied cross and vertical lines on the metal surface, and quality selection of ribbon figured glass bring about abundant facade effect under lighting in different directions and further enhance commercial atmosphere of the shopping center.

03 广场主入口
04 广场外立面

HAKKA YU 客語
watsons 屈臣氏
KEYROAD
Cabbeen
JACK JONES
Levi's
501

06

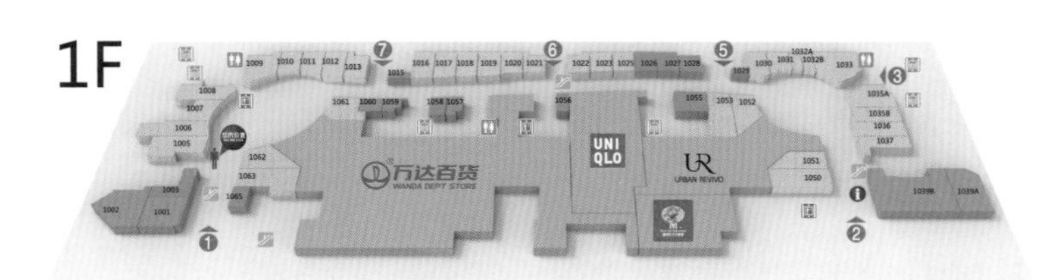

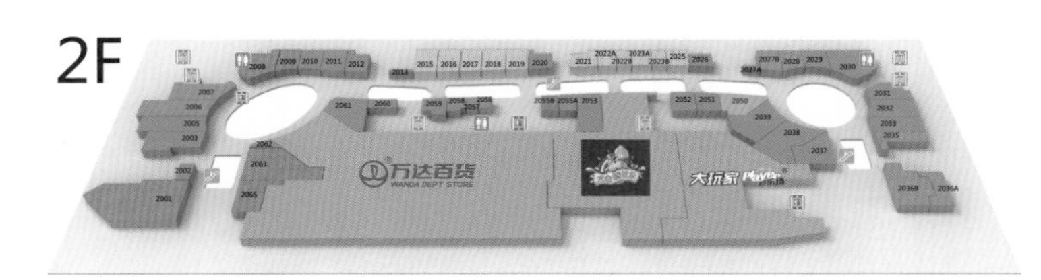

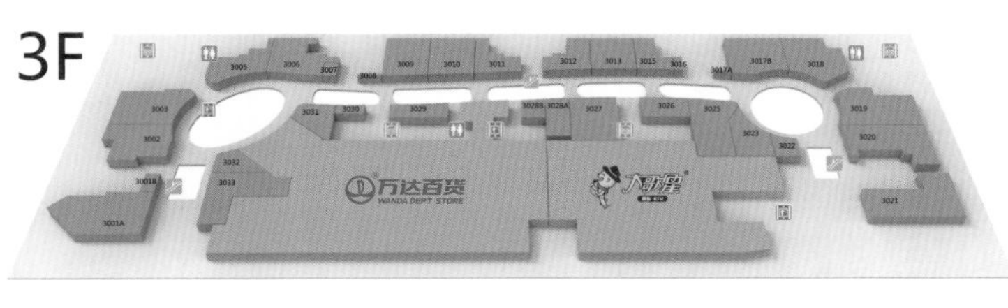

07

08

INTERIOR OF PLAZA
广场内装

中庭以明快轻松的设计手法，结合材质本身的特点，塑造出富有动感的多层空间。以观光电梯为中心向两侧进行塑造，在合理的结构基础上，通过造型的变化使中庭得到强调。灰镜钢材质的运用使商业氛围得到强化，从而达到吸引注意力的设计效果。

The atrium creates a dynamic multi-level space by taking advantage of characteristics of the materials in a lively and breezy design way. The interior design is spread from the sightseeing elevator to both sides. Based on reasonable structure, changing of shapes highlights the atrium. Application of gray mirror finished stainless steel enhances the commercial atmosphere and thus achieving the eye-catching design goal.

05 圆中庭
06 椭圆中庭
07 商铺落位图
08 室内步行街

09

10

11

LANDSCAPE OF PLAZA
广场景观

大商业带状广场由一系列的造型花坛与大商业建筑围合而成。花坛立面呈米白色，水磨石为底，加上细致的马赛克线条，造型简约而不失精致，时尚而富于动感，结合南方特有的热带植物，打造出具有佛山地域特色的种植群落。

Linear plaza of the large commercial area is enclosed by a series of mosaicultures and commercial buildings. Creamy white facade, terrazzo base and delicate mosaic lines endow the mosaiculture with concise but exquisite, fashionable and dynamic shape. Benefiting from this and special tropical plants in South China, plant community with local characteristics of Foshan City is hereby built.

12

NIGHTSCAPE OF PLAZA 广场夜景

09 主题雕塑
10 景观小品
11 景观绿化
12 广场夜景
13 室外步行街
14 室外步行街

EXTERIOR PEDESTRIAN STREET 室外步行街

室外步行街设计充分融入了佛山当地文化，以佛山“功夫”为主题。立面设计采用现代装饰主义与岭南建筑风格结合的手法，穿插“一代宗师”、“功夫咏春”、“宝芝林”等立面片段，使室外步行街洋溢着中国古典文化的气息而不失现代感，彰显出金街独特气质和尊贵品质感。

Design of the exterior pedestrian street is conducted by centering on the theme of Foshan "Kung Fu" and fully integrates distinctive cultures in local area. Modern ornamentation and Lingnan architectural style are combined for the facade design, and facade fragments, such as "The Grand Master", "Kong Fu Wing Chun" and "Bao Zhi Lin" are additionally interspersed in the design, making the exterior pedestrian street filled with Chinese classical cultural atmosphere and a sense of modern times and highlighting unique temperament and noble quality of the golden street.

13

14

11

MA'ANSHAN WANDA PLAZA
马鞍山万达广场

时间 2014 / 09 / 19 **地点** 安徽 / 马鞍山
占地面积 14.34 公顷 **建筑面积** 76.31 万平方米

OPENED ON 19th SEPTEMBER, 2014
LOCATION MA'ANSHAN, ANHUI PROVINCE
LAND AREA 14.34 HECTARES **FLOOR AREA** 763,100M²

OVERVIEW OF PLAZA
广场概述

马鞍山万达广场位于马鞍山市雨山区，地块范围东至泰山路，南至银杏路，西至太白大道，北至钟山路。广场占地14.34公顷，总建筑面积76.31万平方米，由购物中心、室外步行街、甲级写字楼、五星级酒店、SOHO公寓和住宅等业态组成。

Ma'anshan Wanda Plaza is located in Yushan District of Ma'anshan, east to Taishan Road, south to Yinxing Road, west to Taibai Avenue and north to Zhongshan Road. The Plaza covers a land area of 14.34 hectares and overall floor area of 763,100 square meters and consists of shopping center, exterior pedestrian street, grade-A office building, 5-star hotel, SOHO apartment, residence and other formats.

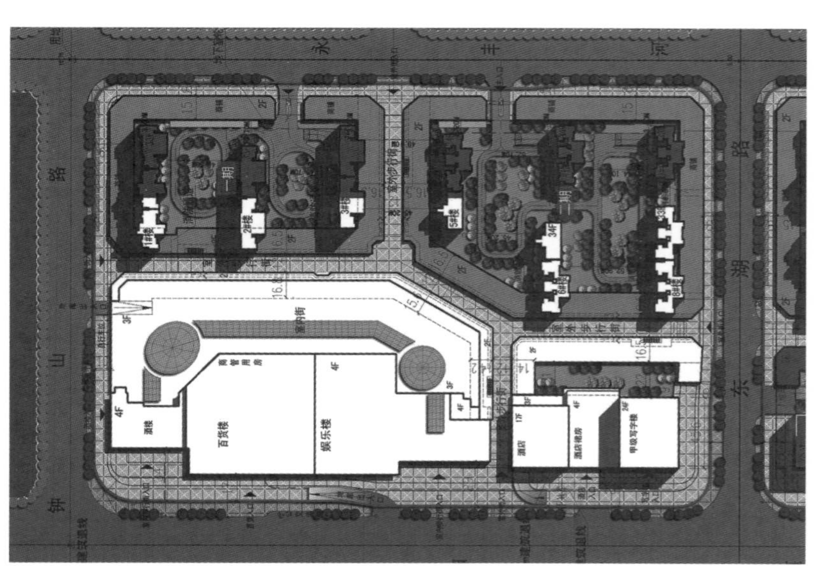

01

01 广场总平面图
02 广场鸟瞰
03 广场立面图

02

03

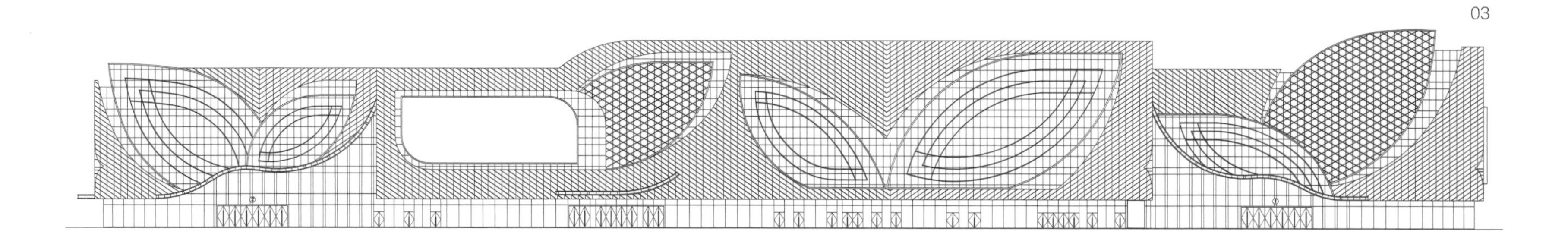

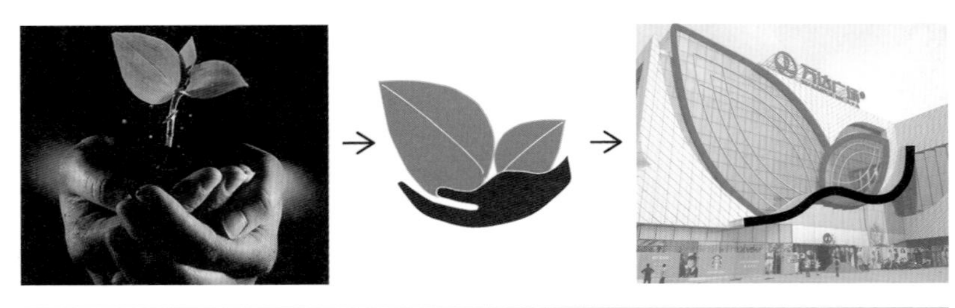

04

05

04 广场外立面“叶片”概念演变图
05 广场外立面“叶片”特写
06 广场外立面局部
07 广场外立面局部
08 广场外立面局部
09 广场外立面

FACADE OF PLAZA
广场外装

马鞍山万达广场外立面以大自然美丽的“叶片”为创意元素，将建筑与自然相融合。在大商业主入口处，自然上扬的“叶片”造型与“双手捧叶”的雨篷相结合，造型立体、凸显曲线的入口雨篷与通透的立面幕墙相结合，突出了大商业主入口，增强了入口的标识性和导向性。用自然叶片脉络抽象出的折线玻璃幕墙，来衬托“叶片”元素，形成了像阳光透过叶片产生的叶脉效果，丰满了整个沿街立面。建筑外立面上富有韵律的“叶片”元素与简洁的形体有机结合，形成了方整大气、特点鲜明的建筑形象。

Facade of the Ma'anshan Wanda Plaza utilizes the creative element of beautiful “leaves” to integrate building with the nature. At main entrance of the large commercial area, the flying “leaf” style combined with canopy in the shape of “leaf in hands” and stereo and curved canopy at the entrance against the background of translucent curtain wall highlight main entrance of the large commercial area and strengthens identifying and guiding role of the entrance. Folded glass curtain wall, extracted from veins of leaves, is utilized to set off the “leaf” element, creating an effect of leaf vein as if sunshine shines through leaves and enriching the whole frontage facade. Rhythmic “leaf” element on the building facade and the concise shape are organically integrated together, creating a magnificent and distinctive building image.

06

07

08

09

10

11

10 椭圆中庭
11 室内步行街
12 商铺落位图
13 圆中庭

1F

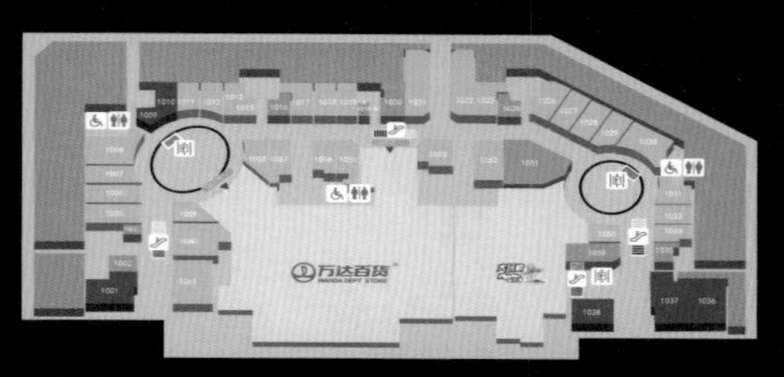

2F

3F

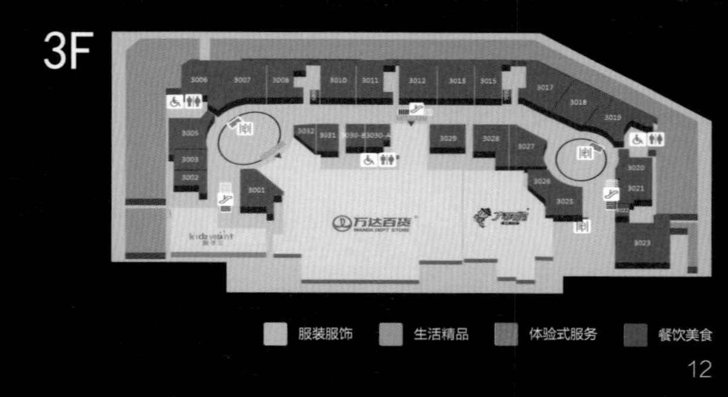

12

INTERIOR OF PLAZA
广场内装

内装设计延伸了"叶片"这一造型元素。室内步行街以树林为主题展开，将长街设计成林荫小路，连桥设计成扭转的折线造型，像树丛中交错的枝杈。椭圆中庭部分，希望营造一个森林中豁然开朗的空间，侧板由白色GRG材料做成水波纹的褶皱造型，就像水面泛起的白色的涟漪。巨大的异形观光电梯下窄上宽，使用三种反射膜玻璃由下至上渐变，观光电梯像是强壮的树干托起了三角形分割的采光顶。圆形中庭作为项目中另外一个汇集空间，希望能和椭圆中庭有所区别，更像树林中的一片开阔地。在侧板的和地面的处理上与长街接近，侧板使用GRG做成折线形状中间凸起，折线沿着中庭侧板上下波动，使整个中庭具有律动感，同时又增加了一些叶子形状的菱形图案在侧板，使细节更加活跃。

Interior design of the Plaza further develops the modeling element of "leaf". The interior pedestrian street is designed like a tree-lined path centering on the theme of woods, and connecting bridges are designed in torsional folded shape, just like interlaced branches in the woods. The oval atrium is expected to build an open space in the dense forest, where side boards are made of white GRG materials in ripple shapes like while ripples on waters. The huge irregular sightseeing elevator, which is large on top and narrow below, adopts three reflective coated glasses to create a bottom-up gradient effect. Like a powerful tree trunk, the sightseeing elevator lifts up the triangle-shaped skylight. As another collecting zone of the Plaza, the circular atrium is expected to be different from the oval atrium and more like an open space in woods. Processing of side boards and floors of the circular atrium is similar to that of the pedestrian street, as for side boards, GRG material is processed in folded shape with the middle rising up and the folded line fluctuates around side boards of the atrium, endowing the atrium with rhythm. Moreover, some leaf-shaped diamond patterns are added for side boards, making details more vivid and lively.

13

LANDSCAPE OF PLAZA
广场景观

景观设计将地域特色文化“花山夜雨”加以利用，形成马鞍山万达广场特有的环境元素。将立面的叶片造型融入景观设计，如将散落有致的绿叶铺装、高大俊秀的景观雕塑合理地分布于项目的主要节点之中，与建筑外立面相映成趣。

Landscape design of the Plaza takes advantage of distinctive local culture of “Night Rain of Huashan” to build unique environmental elements specific to Ma'anshan Wanda Plaza. Leaf modeling of the facade is also applied in landscape design, for example, scattered greenery pavement and elegant landscape sculptures are reasonably distributed at main nodes of the project, forming delightful contrast with the building facade.

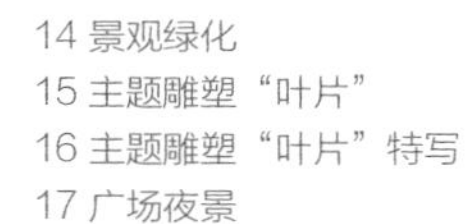

14 景观绿化
15 主题雕塑“叶片”
16 主题雕塑“叶片”特写
17 广场夜景
18 室外步行街

14

15

16

NIGHTSCAPE OF PLAZA
广场夜景

17

EXTERIOR PEDESTRIAN STREET
室外步行街

金街位于万达广场核心位置，贯穿了整个广场，空间上有绵延不绝的气势。设计上着重强化金街连续的空间特质，将大商业的叶片元素转化为“嫩芽”融入金街之中，丰富了立面效果，提升了商业氛围。金街立面融入了红色、橘色等色彩元素，烘托了金街繁荣的商业氛围。构架造型增加了金街天际线的变化，将人们的视线自然而然地引导到二层商铺。立面设计有效地通过视觉效果、空间变化等手段，展现了室外商业步行街的特质。

Located in core position of the Plaza, the golden street runs through the whole plaza, appearing to be endless in spatial effect. Design of the golden street puts emphasis on its spatial quality of continuity, and leaf element adopted in the large commercial area is continually applied in the golden street in the form of “tender shoot” to enrich the facade effect and enhance commercial atmosphere. Red, orange and other elements of color are also integrated into facade of the golden street to highlight its prosperity, and frame shapes add some variation to skyline of the golden street and naturally attract people's views to stores on the first floor. The facade design effectively present qualities of the exterior commercial pedestrian street through visual effect and space variation etc.

18

12

JINGZHOU WANDA PLAZA
荆州万达广场

时间 2014 / 09 / 20 **地点** 湖北 / 荆州
占地面积 12.17 公顷 **建筑面积** 60.75 万平方米

OPENED ON 20th SEPTEMBER, 2014
LOCATION JINGZHOU, HUBEI PROVINCE
LAND AREA 12.17 HECTARES **FLOOR AREA** 607,500M²

OVERVIEW OF PLAZA
广场概述

荆州万达广场项目坐落于荆州市荆州区，位于北京西路与武德路交会的武德片区中心，东接武德路，南临北京西路，西至东门，北达江津西路，占地12.17公顷，总建筑面积60.75万平方米（地上47.55万平方米，地下13.20万平方米）。其中五层购物中心（地上建筑面积8.43万平方米、地下建筑面积7.20万平方米），涵盖了万达百货、万达影城、大歌星、大玩家及超市等多个主力业态。

Jingzhou Wanda Plaza is located at the center of Wude area, intersection of Beijing West Road and Wude Road, in Jingzhou district of Jingzhou city, facing Wude Road to the east, Beijing West Road to the South, the East Gate to the west and Jiangjin West Road to the north. The Plaza covers a land area of 12.17 hectares and overall floor area of 607,500 square meters (including 475,500 square meters for the aboveground part and 132,000 square meters for the underground part). The five-floor shopping center, with aboveground floor area of 84,300 square meters and underground floor area of 72,000 square meters, covers various anchor activities including Wanda Department Store, Wanda Cinema, Superstar, Super Player and supermarket, etc.

01 广场总平面图
02 广场外立面

01

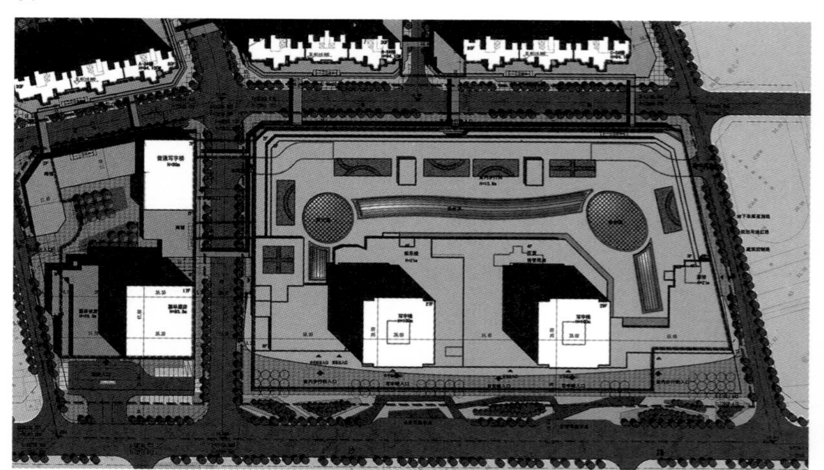

02

FACADE OF PLAZA
广场外装

荆州万达广场立面蜿蜒跌宕的造型，简洁有力的空间形象，给人雕塑般强有力的视觉冲击，相互咬合翻转的三维曲面由3条空间曲线演化而成，整个演化过程由Rhino & Grasshopper参数化设计支撑，在富于动感且舒张的曲面造型背后是逻辑的参数化演算。曲面的造型丰富了建筑形体的轮廓，犹如流动的河流，使得建筑立面具有动态的建筑美感。

Facade of Jingzhou Wanda Plaza enjoys meandering and flowing modeling and simple but powerful space image, giving people sculpture-like visual shock. Its interlocking revolved 3D curved surface is evolved from three space curves and the whole process of evolution is backed up by Rhino & Grasshopper parametric design. That is to say, the dynamic and spreading curved shape is attributed to logical parametric calculation. The curved shape, just like a flowing river, enriches the building profile and endows the building facade with dynamic architectural beauty.

04

05

03

INTERIOR OF PLAZA
广场内装

荆州古城墙作为当地最为显著的地理标志之一，其高超的建筑技艺和不朽的艺术价值，体现了汉民族传统规划思想和建筑风格。内装以城墙体块叠错的造型为出发点，引入时尚现代的流行元素，延展出错落有致的盒子概念，作为整体空间的设计元素。

As one of the most distinctive geographical indications, the ancient city wall is known for its superb building skills and immortal artistic value that embody traditional planning concept and architectural style of Han nationality. Interior design of the Plaza starts with the shape of city wall with stacked blocks and introduces modern fashion elements to develop the conception of stacked boxes as design element of the whole space.

1F

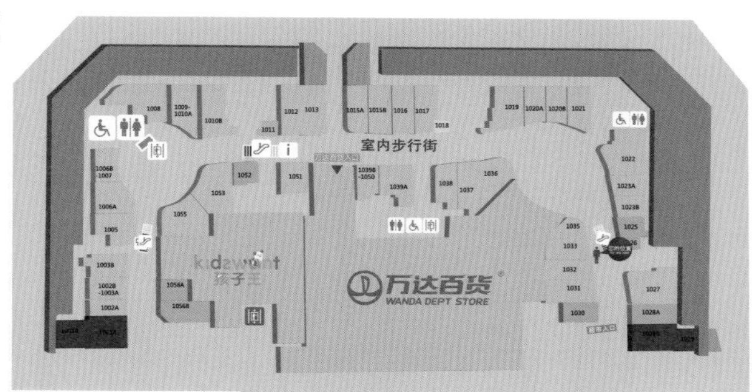

2F

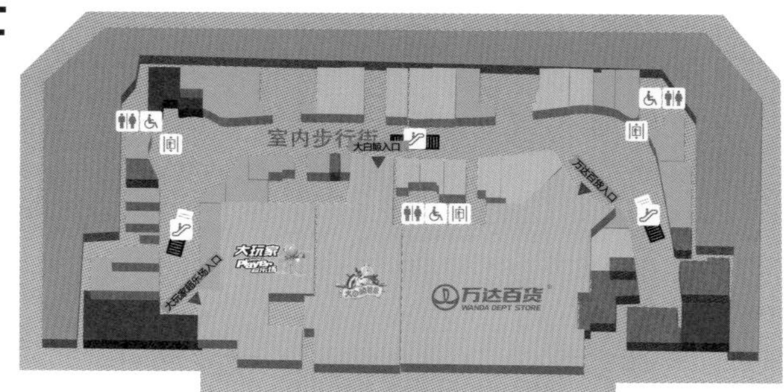

3F

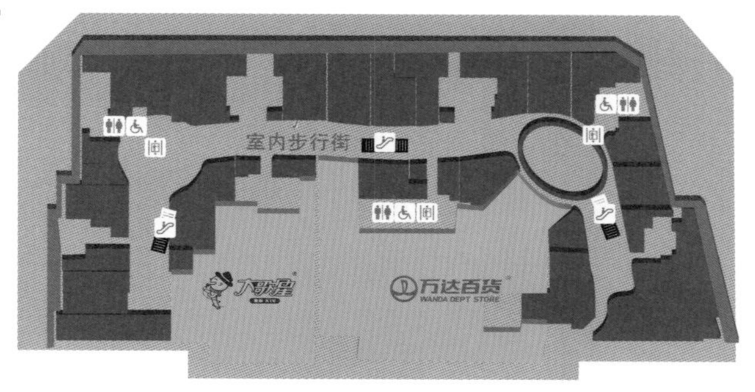

06

服装服饰　生活精品　体验式服务　餐饮美食

03 广场外立面
04 广场外立面局部
05 广场外立面局部
06 商铺落位图
07 圆中庭

07

08

09

10

11

LANDSCAPE OF PLAZA
广场景观

景观总面积约为4.28万平方米，以“鱼水情深”为概念设计主题，在现代时尚的商业空间中融入荆州地域特色文化。主题雕塑位于东侧主广场入口，与水景完美结合。雕塑通过三个在空间上错落的风帆组合在一起，风帆造型简洁，弧形有张力，形似在大海之上乘风破浪。为活跃广场的气氛，增加场地的趣味性，设置造型独特的花钵组合，具有浓郁民俗文化的情景雕塑。雕塑主题取自当地民俗“楚舞”、“采礼”、“采莲船”和“乐”等。

Landscape design of the Plaza, involving a total area of around 42,800 square meters, insists on the conceptual design theme of “Close as Fish and Water” and integrates distinctive local cultures of Jingzhou City into the modern and fashionable commercial space. The theme sculpture, which is composed of three well-spaced sails combined together, is placed at entrance of the east main plaza, echoing with waterscapes. Simple shape and tensile curve of the sail create an image of sailing through the ocean. Moreover, unique flowerpot combination and scene sculptures with strong folk culture are set to activate atmosphere and add fun to the Plaza. Themes of the sculptures are conceived from local folk customs, such as “The Dances of Chu”, “Gifts Procurement” and “Lotus Gathering Boat”, etc.

08 景观小品
09 景观小品
10 景观绿化
11 主题雕塑
12 广场夜景
13 室外步行街

NIGHTSCAPE OF PLAZA
广场夜景

12

EXTERIOR PEDESTRIAN STREET
室外步行街

13

文华酒店 WandaVist

WANDA HOTELS
万达酒店

WANDA
COMMERCIAL
PLANNING
2014

昆明万达文华酒店
烟台万达文华酒店
东莞万达文华酒店
广州增城万达嘉华酒店
潍坊万达铂尔曼酒店
赤峰万达嘉华酒店
济宁万达嘉华酒店
金华万达嘉华酒店
常州万达嘉华酒店

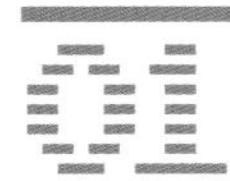

WANDA VISTA KUNMING
昆明万达文华酒店

时间 2014 / 10 / 31 **地点** 云南 / 昆明
建筑面积 4.46 万平方米

OPENED ON 31st OCTOBER, 2014
LOCATION KUNMING, YUNNAN PROVINCE
FLOOR AREA 44,600M²

OVERVIEW OF HOTEL
酒店概况

作为超五星级酒店的昆明万达文华酒店，位于昆明市西山区前兴路，与商业中心、城市步行街、超5A甲级写字楼等共同打造昆明商业地产中的No.1。酒店拥有297间宽敞舒适的客房及套房。餐饮设施齐全，包括“美食汇”全日餐厅、“品珍”中餐厅以及“辣道”川湘料理。酒店的宴会及会议总接待面积接近2200平方米，豪华无柱式大宴会厅面积达1300平方米。

As super 5-star hotel, Wanda Vista Kunming is located at Qianxing Road, Xishan District in Kunming, and jointly builds Number One commercial real estate in Kunming coupled with the commercial center, pedestrian street and super 5A and class A office building. The hotel features 297 capacious and comfortable guestrooms and suites, well-equipped catering facilities including "Café Vista", "Zhen" Chinese Restaurant and Chili & Pepper Specialty Restaurant, about 2,200 square meters total reception area of banquet and meeting are and 1,300 square meters pillar-free luxurious grand banquet hall.

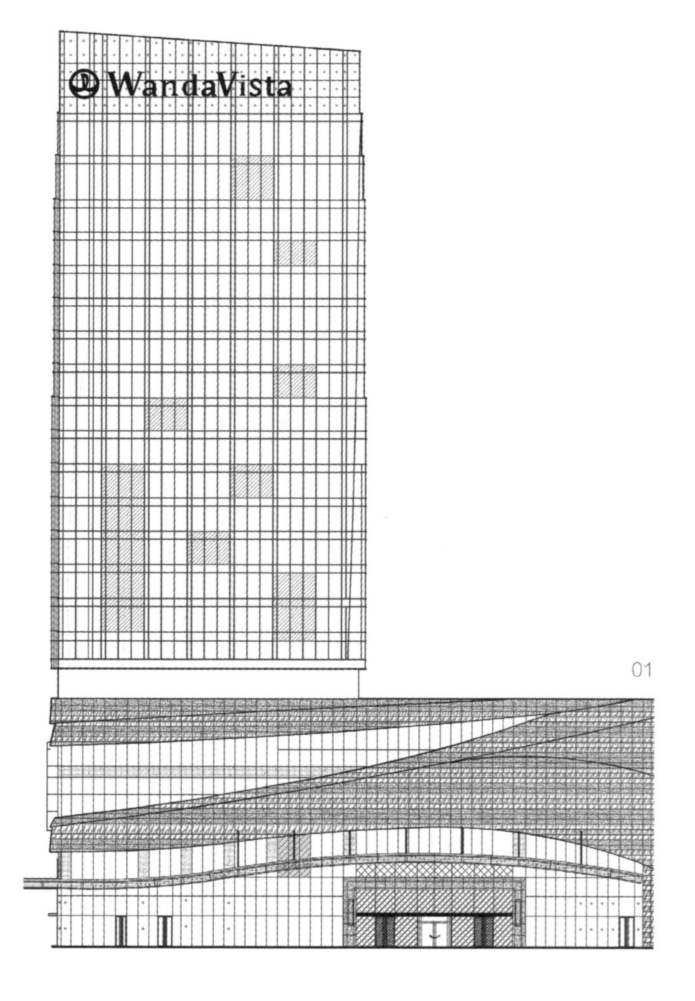

01

02

01 酒店立面图
02 酒店外立面

万达文华酒店 WandaVista
昆明万达文华酒店

FACADE OF HOTEL
酒店外装

酒店外立面方案构思来源于昆明市花山茶花，塔楼由数片形似花瓣的玻璃幕墙拼合而出，体量挺拔，形态饱满，有强烈的视觉冲击力；裙房以石材为主，通过深浅交替的色带自由错动，曲线的玻璃窗户凹进石材幕墙，显得体量感十足、端庄典雅。酒店主入口雨篷及门套造型是整个建筑的“华彩乐章”，采用双曲弧形的造型，底部配以花瓣形式的铝板天花造型，与整个立面弧形相呼应，也增强了入口的装饰感。

The facade design is originated from camellia, the City-flower of Kunming. The tower is built by splicing several petal-like glass curtains, presenting tall and straight volume, rich shape and powerful visual impact; the podium adopts the stone as the main material and through free movement of alternative dark or light colored tapes and the insertion of curvilinear glass window into stone curtain wall, consequently shows elegant and dignified building with a strong dimension sense. As “the colorful chapter” of the whole building, the hotel's main entrance canopy and door pocket modeling adopt the hyperbolic arc shape, whose bottom is decorated with aluminum ceiling shaped by petal, thus echoing with arc-shaped facade and also enhancing the decorative perception of entrances.

04

03 酒店入口
04 酒店外立面

INTERIOR OF HOTEL
酒店内装

05

LANDSCAPE OF HOTEL
酒店景观

大型壁画《古滇之歌》吸取了石寨山青铜文化的各种造型特点及精巧的构成语言：青铜与金铂相结合、人与动物的写实艺术、人物夸张的舞蹈、地方风格的装饰造型、古滇建筑形态和古代云南土著居民日常生活场景等元素，充分体现了云南地域、历史的特色。

The large-scale fresco "Song of Ancient Dian" is inspired from the features of various modeling and delicate design language contained in bronze culture of Shizhai Mountain: the combination of bronze and gold platinum, realism arts of human and animal, exaggerated dance, decorative model with local style, building forms in Ancient Dian and daily living scenes of ancient Yunnan aborigines. The hotel design fully demonstrates the regional and historical features.

05 酒店大堂
06 酒店内院水景

06

07

07 酒店景观墙
08 屋顶景观亭
09 内院景观小品
10 酒店入口夜景

08

09

NIGHTSCAPE OF HOTEL
酒店夜景

10

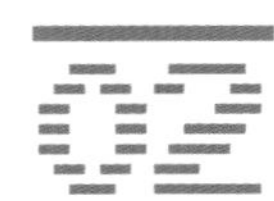

WANDA VISTA YANTAI
烟台万达文华酒店

时间 2014 / 11 / 21 **地点** 山东 / 烟台
建筑面积 4.4 万平方米

OPENED ON 21st NOVEMBER, 2014
LOCATION YANTAI, SHANDONG PROVINCE
FLOOR AREA 44,000M²

OVERVIEW OF HOTEL
酒店概况

烟台万达文华酒店坐落于烟台市中心的芝罘区，北侧毗邻烟台市文化中心，地理位置优越，交通便利。酒店位于由两栋200米超高层组成的双塔（北塔）上部，是万达第一个位于超高层写字楼上方的酒店，建筑规模4.4万平方米。地上两层裙房拥有1400平方米的宴会大厅及各种类型会议厅供顾客选择。客人进入首层到达大堂后，通过酒店穿梭电梯可直达26层空中大堂，尽揽海景；27层为海景全日餐厅，28层为中餐厅，29层为康体娱乐区。酒店客房共有307间，分布在30层-41层，而位于最高楼层的行政酒廊和总统套更是坐拥整座城市的至高点。

Located in Zhifu District, center of Yantai City and north of Cultural Centre in Yantai, Wanda Vista Yantai enjoys an advantageous geographical location and convenient transportation. The Hotel is positioned at top of the North of landmark twin towers, composed of 200 meter super high-rise buildings, the highest in Yantai, became the first Wanda hotel that is positioned on the top of super high-rise towers. Building scale of this project reaches 44,000 square meters, and the two-floor podiums aboveground is arranged with 1,400 square meters banquet hall and meeting rooms in various sizes and types. After entering the ground floor lobby, the hotel express elevators deliver visitors direct at the 25F sky lobby to appreciate the ocean panorama. 24h Harbor View Dining, Chinese restaurant and entertainment area are respectively located on the 26F, 27 F and 28F, all have splendid ocean views. There are 307 rooms in total distributed on the 29F to 40F and executive lounge and presidential suite on the top floor of the Hotel stand at the highest point of Yantai.

01

01 酒店远景
02 酒店近景

02

03

FACADE OF HOTEL
酒店外装

建筑立面设计的灵感来源于广阔的海洋和烟台深厚的历史文化——碧海无边，烟气浩渺，以“碧海云烟”为主题，突出时尚、简约、明快的建筑风格。蓝灰色的玻璃幕墙上，搭配白色铝板线条节节延伸，有如天空中云朵的倒影在大海。高雅的艺术格调与内涵完美融合，诠释了现代简约的浪漫情怀，给人以优雅舒适的观感体验。

Inspired by the vast ocean and long historic culture of Yantai, saying boundless blue sea and vast mist-covered waters, facade design of the Hotel is centered in the theme of "Sea and Clouds" and highlights the fashionable, simple and lively architectural style. The blue grey glass curtain walls are spread with white aluminum plate moldings, refernce to the image of clouds in the sky reflected on the ocean. Elegant artistic style and connotation are perfectly integrated to interpret modern and pure romantic theme and render an elegant and comfortable experience.

03 酒店入口
04 酒店大堂

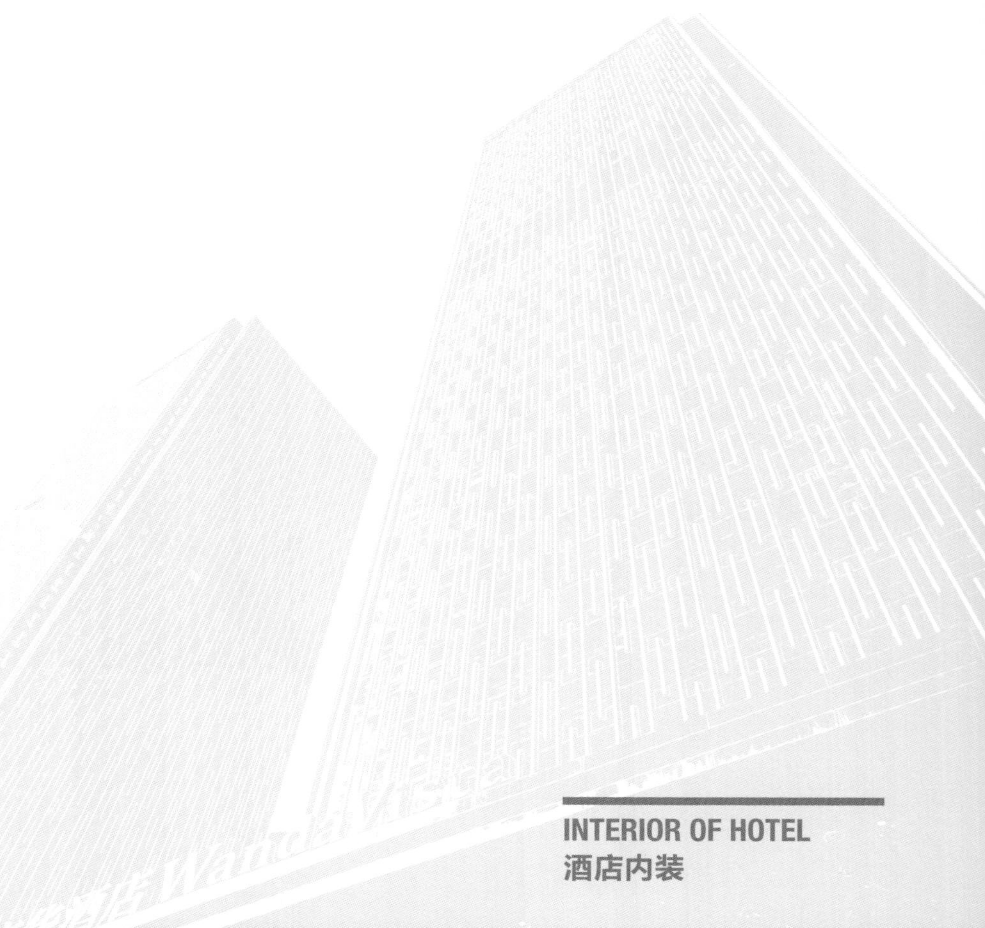

INTERIOR OF HOTEL
酒店内装

04

LANDSCAPE OF HOTEL
酒店景观

烟台酒店旨在打造秀丽海岸风情，挖掘项目地域特色，彰显地域特征。设计提取半岛风情特色元素，与酒店奢华大气的风格合而为一，结合场地功能交错的特点，合理疏导流线，更好地营造酒店空间应用。从水景到绿植，从铺装到构筑，从色调到材质，注重细节打造的同时，更注重驾驭整体空间的感受，构建出一个功能齐全，大气奢华的酒店景观氛围。

With exploration into and demonstration of its regional characteristics of coastal beauty, Wanda Vista Yantai is built with a distinctive feature and style. The design combines the characteristic elements of peninsula feature and the hotel's luxurious and elegant style into one. In combination with the hotel's intertwined functions and reasonable planning and outlining for streamline, a better hotel layout is achieved. With special care given to details, from waterscape to green plants, from pavement to construction, from color to texture, every single detail of the design is delicately portrayed. Meanwhile, with attention given to the sense of the overall space, the hotel delivers a luxurious and elegant landscape atmosphere with perfected functions.

05 景观绿化
06 酒店入口水景
07 酒店夜景

06

NIGHTSCAPE OF HOTEL
酒店夜景

07

WANDA VISTA DONGGUAN
东莞万达文华酒店

时间 2014 / 09 / 12 **地点** 广东 / 东莞
建筑面积 4.45 万平方米

OPENED ON 12th SEPTEMBER, 2014
LOCATION DONGGUAN, GUANGDONG PROVINCE
FLOOR AREA 44,500M^2

OVERVIEW OF HOTEL
酒店概况

东莞万达文华酒店坐落于东莞市东城区中心商务核心区，毗邻东莞东城万达广场，与著名的旗峰山咫尺之遥，交通便利；酒店拥有306间宽敞舒适的客房及套房，设计高雅、配套豪华。餐饮设施齐全，包括“美食汇”全日餐厅、“品珍”中餐厅以及“和”日料餐厅。酒店的宴会设施包括可容纳近1000人同时用餐的大宴会厅以及东莞市唯一的超100平方米的LED屏幕。

Wanda Vista Dongguan has convenient transportation right at CBD of Dongcheng District, Dongguan City and is adjacent to Dongguan Dongcheng Wanda Plaza and the famous Qifeng Mountain. With 306 capacious and comfortable rooms and suites in total, the Hotel is known for its elegant design, luxury facilities, complete dining facilities including all day dining “Café Vista”, “Zhen” Chinese Restaurant and Japanese restaurant “Yamato”, and banquet facilities including large banquet hall that can accommodate 1000 persons for dining at the same time and the sole LED screen over 100 square meters in Dongguan.

01

01 酒店远景
02 酒店近景

02

03

03 酒店入口
04 酒店大堂

FACADE OF HOTEL
酒店外装

酒店利用垂直元素强调建筑的竖向挺拔感，旨在体现纯净和现代的立面效果。建筑使用幕墙体系，垂直元素坚实明亮，富含节奏韵律。雨篷采用彩釉花格玻璃，华丽端庄中又饱含典雅细腻。材料方面，酒店使用珍珠白的压花玻璃，写字楼使用铝板百页条带，并使用了三种不同玻璃，使得细节得到更多展示，体现了更加丰富的材料质感。

The Hotel takes advantage of vertical elements to strengthen the tallness and straightness of the building, aiming to show a pure and modern facade effect. The building selects curtain wall system with solid and bright vertical elements, showing a sense of rhythm, and canopy adopts enameled lattice glasses, gorgeous but graceful. In terms of materials, pearl white figured glass and aluminum plate louver strips are respectively adopted for the hotel and office building. Moreover, three different glasses are also applied to present more details and stronger quality sense of materials.

INTERIOR OF HOTEL
酒店内装

04

LANDSCAPE OF HOTEL
酒店景观

景观结合并延续建筑竖线条的肌理，从珠光水韵中提取的圆形元素结合构成了主要的表现形式。通过软硬景的结合来烘托建筑，使其生命力更加旺盛，入口有气势的水景更彰显酒店的高端大气。酒店后场粗糙的石材质感与静谧的水池形成鲜明的对比，既烘托了酒店高端安静的氛围，同时清澈的跌水与明珠也交相呼应。

Landscape design of the hotel combines and extends the building texture of vertical lines and forms the main representation constituted by circular elements extracted from pearls and waters. Backed up by combination of soft and hard landscapes, the building appears to be more vivid and vital. Moreover, grand water features at the entrance further manifest grandness and magnificence of the Hotel, rough stone materials and quiet pools at the back of house in sharp contrast strengthen quietness and peace of the Hotel, and the crystal clear water drops also echoes the pearls.

05 酒店入口景观
06 酒店入口水景

05

06

07

08

NIGHTSCAPE OF HOTEL
酒店夜景

09

07 酒店绿化
08 酒店屋顶花园
09 酒店夜景

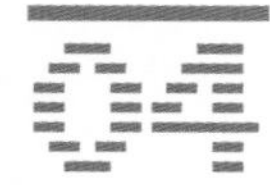

WANDA REALM GUANGZHOU ZENGCHENG
广州增城万达嘉华酒店

时间 2014 / 05 / 16 **地点** 广东 / 广州
建筑面积 3.64 万平方米

OPENED ON 16th MAY, 2014
LOCATION GUANGZHOU, GUANGDONG PROVINCE
FLOOR AREA 36,400M^2

OVERVIEW OF HOTEL
酒店概况

广州增城万达嘉华酒店位于广州增城区增景大道与荔城大道交口处，总建筑面积3.64万平方米，拥有客房286套。酒店主体建筑地上17层，建筑面积3.04万平方米；地下2层，建筑面积0.6万平方米。酒店首层大堂面积700平方米，挑空层高13米，设置大堂吧及全日餐厅等；二层为中餐厅及特色餐厅；三层为宴会、会议层，包括一个面积为1200平方米的大宴会厅及多个大小会议室；四层为康体层，包括健身房、游泳池等康体设施；客房楼顶部的行政楼层包含高级套房、行政酒廊以及总统套房。

Wanda Realm Guangzhou Zengcheng is located at the intersection of Zengjing Road and Licheng Road in Zengcheng District of Guangzhou, covers an overall floor area of 36,400 square meters and includes 286 guest rooms. Main building of the Hotel has 17 floors aboveground with a floor area of 30,400 square meters and two floors underground with a floor area of 6,000 square meters. The ground floor is arranged with a lobby with area of 700 square meters and raised floor height of 13m, including lobby bar and all day dining; the first floor is designed with Chinese restaurant and Specialty restaurant; the second floor is the banquet and meeting floor including one 1,200 square meters large banquet hall and several meeting rooms in various standards; the third floor is the fitness and recreation floor including gym, swimming pool and other sports and recreation facilities, and the executive floor on top of the guest room building includes superior suite, executive lounge and presidential suite.

01 酒店外立面

FACADE OF HOTEL
酒店外装

酒店外立面强调竖向，并采用浅色调以减弱建筑体块对视觉的冲击，楼层间玻璃倾斜拼接，从而丰富立面造型，增强建筑空间感受。竖向铝材凸出墙面的处理手法，突显出建筑主体挺拔且富张力，主体与裙房顶部采用收缩处理，更加使得建筑极具活力，同时也增强了建筑空间层次感。

Facade of the Hotel emphasizes vertical design, where light colors are adopted to weaken visual impact caused by building blocks and glasses between floors are spliced in an inclined way to enrich the facade shape and strengthen the sense of space of the building. Vertical aluminum materials are processed to be projected from the walls, thus to highlight a straight, upright and tensile main building. In addition, the main building and podiums top adopt contraction process to make the building vivid and dynamic and strengthen space layering of the building.

INTERIOR OF HOTEL
酒店内装

03

NIGHTSCAPE OF HOTEL
酒店夜景

04

02 酒店入口
03 酒店大堂
04 酒店夜景

WANDA PULLMAN WEIFANG
潍坊万达铂尔曼酒店

时间 2014 / 05 / 23 **地点** 山东 / 潍坊
建筑面积 3.68 万平方米

OPENED ON 23rd MAY, 2014
LOCATION WEIFANG, SHANDONG PROVINCE
FLOOR AREA 36,800M²

01 酒店入口
02 酒店外立面

OVERVIEW OF HOTEL
酒店概况

潍坊万达铂尔曼酒店位于潍坊市奎文区，酒店总建筑面积3.68万平方米。东部客房可将虞河公园尽收眼底，饱览美丽景色。酒店三层设置1100平方米的大宴会厅，会议区配置5个中型会议室。酒店内豪华水疗、健身及室内泳池一应俱全，泳池外配有屋顶活动平台，为宾客提供阳光浴空间，充分享受五星级酒店的超值服务。

Wanda Pullman Weifang is located in Kuiwen District of Weifang and has an overall floor area of 36,800 square meters. Guests in guest rooms on the east side can overlook the Yuhe River Park and enjoy a panoramic view of beautiful scenery. The Hotel is completed with a 1,100 square meters large banquet hall set on the second floor and five medium-sized meeting rooms arranged in the meeting area. Besides, luxury Spa, gym and indoor swimming pool that is added with roof activity platform to provide visitors a space for sunshine are all available in the Hotel, enabling guests to enjoy premium services of a five-star hotel.

01

02

03

FACADE OF HOTEL
酒店外装

酒店整体的立面风格洋溢着简约典雅的时尚气质，贯穿全身的竖向线条优雅修长，充满向上的力量。竖向线条在裙房和塔楼顶部做宽窄的节奏变化，形成简洁明快的韵律，整体效果典雅大气。在草长莺飞，柳醉春烟的城市中，万达铂尔曼酒店就像在天空飞翔的一只美丽风筝，成为潍坊城市的个性地标。酒店雨篷出挑深远，大气华丽，构思来源于风筝的龙骨。其底部的造型犹如风筝的龙骨扎制而成，主次分明，体现出民间艺术之美，富有潍坊的地方特色。

Overall style of the Hotel facade appears to be concise and elegant, and the graceful and long vertical lines spread over the facade present a positive upward power. Vertical lines are processed with rhythm changes in width to form a simple but vivid rhythm, exhibiting grand and elegant overall effect. Set in the city full of life, Wanda Pullman Weifang, like a kite flying in the sky, becomes a special landmark of Weifang City. Conceived from keels of the kite, canopy of the Hotel is deeply overhung, grand and magnificent, and the bottom shape appears to be bound by kite keels and well-organized, showing beauty of folk arts and local features of Weifang.

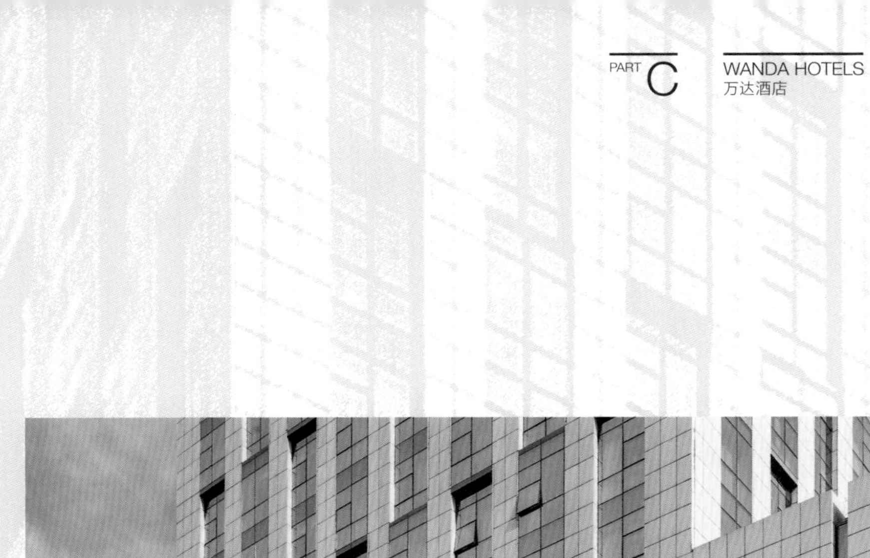

04

INTERIOR OF HOTEL
酒店内装

05

03 酒店外立面
04 酒店入口
05 酒店大堂

LANDSCAPE OF HOTEL
酒店景观

潍坊酒店的景观营造了静谧、优美的氛围——酒店与外界有绿树浓荫隔离，内庭院通过错落叠水、竹林、疏林草地等精美景观元素创造出清净、高雅的景观空间。

Boasting of green trees to isolate itself from the outside world, and peaceful and elegant inner court landscape space forged by exquisite landscape elements such as staggered cascading, bamboo forest, trees and lawn, the whole hotel has its landscape immersed in a tranquil and graceful atmosphere.

06

07

08

NIGHTSCAPE OF HOTEL
酒店夜景

09

06 酒店水景
07 景观绿化
08 酒店花园
09 酒店夜景

WANDA REALM CHIFENG
赤峰万达嘉华酒店

时间 2014 / 06 / 20 **地点** 内蒙古 / 赤峰
建筑面积 4.7 万平方米

OPENED ON 20th JUNE, 2014
LOCATION CHIFENG, INNER MONGOLIA AUTONOMOUS REGION
FLOOR AREA 47,000M²

OVERVIEW OF HOTEL
酒店概况

赤峰万达嘉华酒店位于赤峰市红山区，总建筑面积4.7万平方米，其中：主体建筑地上20层，建筑面积4万平方米；地下2层，建筑面积0.7万平方米。酒店设客房350套，拥有五星级酒店相应的配套设施。其中，一层为酒店大堂、大堂吧、全日制餐厅；二层为特色餐厅、中餐厅及相应包房；三层为会见厅、会议洽谈及宴会厅；四层为游泳区、健身区和行政办公；五至十七层为标准客房层；十八层为标准客房及部长套房区；十九层为标准客房及行政走廊；二十层为标准客房及总统套房区。

Wanda Realm Chifeng is located in Hongshan District of Chifeng, Inner Mongolia Autonomous Region, covers an overall floor area of 47,000 square meters and includes 350 guest rooms. Main building of the Hotel has 20 floors aboveground with a floor area of 40,000 square meters and two floors underground with a floor area of 7,000 square meters. The Hotel is completed with supporting facilities of a five-star hotel, for example, the ground floor is arranged with lobby, lobby bar and all day dining, the first floor is designed with Specialty restaurant, Chinese restaurant and private rooms, the second floor is se with presence chamber, meeting room and banquet hall, the third floor includes swimming, gym and executive office area, from 4F to 16F are standard guest room floors, 17F is the standard guest room and ministerial suite area, 18F is the standard guest room and executive lounge and 19F is the standard guest room and presidential suite area.

01

02

FACADE OF HOTEL
酒店外装

暖黄色的石材构成典雅、挺拔的竖向肌理，显露强烈的传统历史痕迹与浑厚的文化底蕴；同时又摒弃了过于复杂构造和装饰，体现了五星级酒店的奢华与厚重。

Facade design of the Hotel selects stone materials in warm yellow to form an elegant and upright vertical texture, attempting to display strong sense of traditional history and profound cultural deposits. Meanwhile, overly complicated structure and ornaments are abandoned to represent luxury and decorousness of a five-star hotel.

01 酒店外立面
02 酒店入口
03 酒店立面图
04 酒店大堂

INTERIOR OF HOTEL
酒店内装

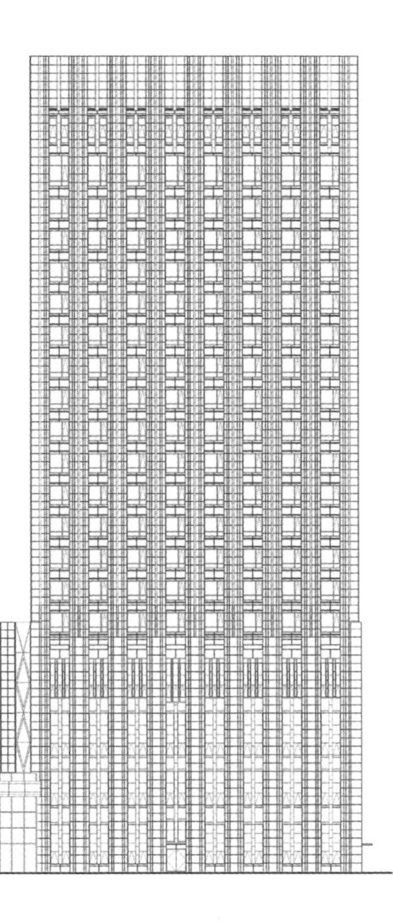
03

04

LANDSCAPE OF HOTEL
酒店景观

项目结合地域特色，如湖泊、山峰、河流等自然形态，将其抽象的自然元素提炼出来：湖泊的“圆形曲线”元素作为景观节点，山峰的“三角”元素作为绿化和地形的塑造，河流的“线性”元素作为景观的轴线；同时将提炼出来的元素结合建筑的直线元素运用于酒店的灯具、铺装和构筑物中。

Regional features of Chifeng, such as lake, peak, river and other natural features are integrated and abstract natural elements are extracted in landscape design of the Hotel, for example, “circular curve” element of lakes are utilized as the landscape nodes, “triangle” element of peaks as molding of greening and terrain and “linear” element of rivers as axis of landscape. Meanwhile, the extracted elements, combined with straight element of the building, are applied in lighting fixtures, pavement and structures of the Hotel.

05 酒店入口水景
06 景观亭
07 酒店夜景

05

06

NIGHTSCAPE OF HOTEL
酒店夜景

07

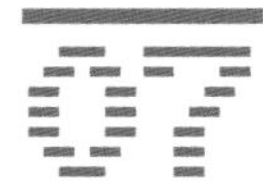

WANDA REALM JINING
济宁万达嘉华酒店

时间 2014 / 07 / 05　**地点** 山东 / 济宁
建筑面积 3.65 万平方米

OPENED ON 5th JULY, 2014
LOCATION JINING, SHANDONG PROVINCE
FLOOR AREA 36,500M²

01 酒店入口
02 酒店外立面

OVERVIEW OF HOTEL
酒店概况

济宁万达嘉华酒店位于济宁市太白东路，建筑面积3.65万平方米，地上17层，客房286间。首层为700平方米挑空大堂、570平方米全日餐厅、210平方米大堂吧及200平方米的精品店；二层为中餐和特色餐，分设包间和散座区；三层设有会见厅、会议室和1200平方米的宴会厅；四层是酒店的康体设施，含游泳池、健身跳操、美容美发等功能。塔楼为客房层，含标准客房、套房、部长套房、总统套房等客房类型，16层为行政酒廊。

Wanda Realm Jining is located on the Taibai East Road of Jining, covers an overall floor area of 36,500 square meters including 17 floors aboveground and has 286 guest rooms. The ground floor is arranged with a raised lobby with area of 700 square meters, a 570 square meters all day dinning, one 210 square meters lobby bar and one 200 square meters boutique; the first floor is designed with Chinese restaurant and Specialty restaurant that are set with private rooms and extra seats; the second floor is set with presence chamber, meeting room and 1,200 square meters banquet hall, and the third floor is set with recreation facilities supporting swimming pool, fitness, beauty salon and other recreation functions. The tower building is designed as guest room floor, including rooms in various standards such as standard room, suite, ministerial suite and presidential suite, and 15F is the executive lounge.

02

FACADE OF HOTEL
酒店外装

外立面采用中式装饰花格图案，在保证外立面效果的前提下，既避免了对客房内视线造成遮挡的问题，又节约了建造成本；同时与雨篷天花、壁灯及酒店内装图案形成呼应。酒店入口通过“门”形铝板构架进行强化。石材门套高度接近大堂天花，内外统一，显得挺拔大气。雨篷天花的装饰花格继续延续了外立面的中式装饰花格主题，强化了文化意味。

Taking advantages of Chinese style lattice pattern, facade design of the Hotel, based on the premise of facade effect, not only avoids blocked sight in guest rooms but also saves the construction cost, it also echoes canopy ceiling, wall lamps and interior finishing patterns. Hotel entrance is highlighted by inverted U-shaped aluminum plate frame and stone door frame, which is as high as the lobby ceiling, forms a unified effect of indoor and outdoor spaces, upright and grand. Decorative patterns of the canopy ceiling follow the theme of building facade to strengthen the cultural meaning.

03 酒店入口
04 酒店入口雨篷
05 酒店大堂

04

INTERIOR OF HOTEL
酒店内装

05

07

08

LANDSCAPE OF HOTEL
酒店景观

酒店景观设计较为简洁——前场设计了气场强大的前场水景，左右绿化使用了大株银杏和白桦；后场根据气候特点设计了彩色浅池底水景，符合四季变换的需求。

Landscape design of the Hotel is quite concise and simple. In the front court area, grand and magnificent water features, with gingko and white birch are adopted for greening on both sides; in the backyard garden, waterscapes are set at colored shallow reflecting pools according to the climatic features to echo four seasons.

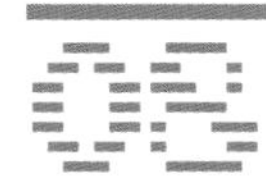

WANDA REALM JINHUA
金华万达嘉华酒店

时间 2014 / 07 / 25 **地点** 浙江 / 金华
建筑面积 4.15 万平方米

OPENED ON 25th JULY, 2014
LOCATION JINHUA, ZHEJIANG PROVINCE
FLOOR AREA 41,500M²

01

OVERVIEW OF HOTEL
项目概况

金华万达嘉华酒店位于金华市金东区，是金华第一家国际标准五星级酒店，酒店主体建筑19层，地上建筑面积3.5万平方米，地下室配套用房0.65万平方米，总建筑面积超过4万平方米，设有各类高档客房332间，中餐厅、全日制西餐厅、特色餐厅一应俱全。酒店还设有游泳池、健身中心、商务中心、会议中心、能容纳60桌1200平方米多功能无柱宴会厅等配套设施。

Located on Jindong District of Jinhua, Wanda Realm Jinhua is the first international five-star hotel in the city. Main building of the Hotel has 19 floors aboveground with floor area of 35,000 square meters and supporting rooms underground with area of 6,500 square meters, and the gross floor area exceeds 40,000 square meters. The Hotel has 332 high-grade rooms in various standards and completed functions including Chinese restaurant, 24h western restaurant, specialty restaurant, swimming pool, fitness center, business center, conference center, 1,200 square meters column-free multifunctional banquet hall that can accommodate 60 tables and other supporting facilities.

02

01 酒店设计手绘稿
02 酒店入口全景
03 酒店外立面

03

FACADE OF HOTEL
酒店外装

酒店设计着重强调了塔楼垂直通透的效果，垂直向上的线条从视觉上使得原有并不高耸的建筑体量显得高耸而挺拔。白色铝板以及深色玻璃一虚一实，相互交映，又使得建筑显得优雅而修长。

Facade design of the Hotel puts emphasis on verticality and transparency of the tower building, vertically upward lines make the ordinary building volume appears to be visually tall and upright. Intersection of the invisible white aluminum plates and visible dark glasses contribute to a tall and graceful building appearance.

04

裙房、雨篷的设计是整个立面的点睛之笔，白色铝板与深色玻璃组成棱锥式的立面形成强烈对比，仿若嵌入建筑的一颗颗宝石，熠熠生辉，优雅而高贵，带动了整个建筑的氛围。挑出的雨篷，如大气华盖，以简洁的横线凹凸收边，勾勒出精致外型。雨篷底部运用天然云石以菱形钻石为主体的造型，与裙房遥相呼应，白天漫射阳光，夜晚配合灯光，闪动剔透晶莹。

Punch line of the whole facade should be the podium and canopy design. The pyramid facade composed of white aluminum plates and dark glasses in sharp contrast looks to be noble and elegant, shining and sparkling, just like jewels inlaid in the building, making the whole building to be more vivid and dynamic. As for overhung canopy of the building that looks like gorgeous canopy over an imperial carriage, simple transverse lines are utilized for edge processing in a concave-convex form to create a delicate shape, and the canopy bottom adopts diamond-shaped natural marbles that echoes podiums at a distance to diffuse sunshine in the daytime and match with lighting as night falls, appearing to be glittering and translucent.

04 酒店入口
05 酒店外立面特写
06 酒店大堂

05

INTERIOR OF HOTEL
酒店内装

06

07

LANDSCAPE OF HOTEL
酒店景观

景观之美不仅体现在造型创意和色彩语言上，心理感受对于客户来说往往更加重要。金华万达嘉华酒店景观设计强调环境和心理学结合，通过植物的精心布局创造可以静思冥想空间，软景和硬质构架的合理组合创造静态休闲空间，室内外空间的高度结合，不同形式的水形态的介入创造出精致的休憩场所。

For beauty of landscape, shape creativity or color language is not the sole influence factor, psychological feeling always means more for guests. Landscape design of the Hotel stresses a combination of environment and psychology and aims to create a space for thinking and meditation by well-arranged plants, build a static leisure space through reasonable organization of soft landscapes and hard frames, reach high integration of interior and exterior spaces and create a delicate open space by introducing waters in different forms.

08

07 景观绿化
08 酒店水景
09 景观绿化
10 酒店夜景

09

NIGHTSCAPE OF HOTEL
酒店夜景

10

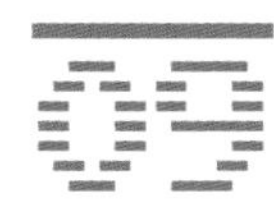

WANDA REALM CHANGZHOU
常州万达嘉华酒店

时间 2014 / 08 / 08 **地点** 江苏 / 常州
建筑面积 3.40 万平方米

OPENED ON 8th AUGUST, 2014
LOCATION CHANGZHOU, JIANGSU PROVINCE
FLOOR AREA 34,000M²

01 酒店入口
02 酒店立面图
03 酒店外立面

OVERVIEW OF HOTEL
酒店概况

常州万达嘉华酒店位于常州市武进区花园街与人民路交叉口，总建筑面积3.71万平方米，其中：主体建筑地上16层，建筑面积2.85万平方米；地下2层，建筑面积0.55万平方米。酒店设客房265套，拥有五星级酒店相应的配套设施，其中：首层大堂面积800平方米，挑空层高13米，还设置大堂吧、全日餐厅及特色餐厅等；二层为中餐厅；三层为宴会、会议层，包括一个面积1200平方米的大宴会厅及多个大小会议室；四层为康体层，包括健身房、游泳池等康体设施；客房楼顶部的行政楼层包含高级套房、行政酒廊以及总统套房。

Wanda Realm Changzhou is located at the intersection of Huayuan Street and Renmin Road, Wujin District, Changzhou, covers an overall floor area of 37,100 square meters and includes 265 guest rooms. Main building of the Hotel has 16 floors aboveground with a floor area of 28,500 square meters and two floors underground with a floor area of 5,500 square meters. The Hotel is completed with supporting facilities of a five-star hotel, for example, the ground floor is arranged with a lobby with area of 800 square meters and raised floor height of 13m, lobby bar, all day dining and specialty restaurant, the first floor is designed with Chinese restaurant, the second floor is the banquet and meeting floor including one 1,200 square meters large banquet hall and several meeting rooms in various standards, the third floor is the recreation floor including gym, swimming pool and other recreation facilities, and the executive floor on top of the guest room building includes superior suite, executive lounge and presidential suite.

01

02

WANDA REALM

万达广场 WANDA PLAZA

WANDA REALM
安全乐四季
健康福百年
WANDA REALM

FACADE OF HOTEL
酒店外装

中华民族以龙图腾为尊，设计取常州地区 “中吴龙城”之名号，运用现代设计材料和语言隐喻了龙的概念。酒店、写字楼、SOHO三种业态由“龙鳞”的元素统一成整体，同时各自单体又另以龙的不同身体部位为元素，组成一个“龙”的形象的建筑群。酒店建筑立面使用了大面积的玻璃，使其在四栋建筑中形象最为特殊，作为“龙头”。酒店晶莹剔透，居整个群落的主导地位。贯穿建筑通体的折面凸窗，凹凸相邻，奇偶错落，鳞次栉比，营造出极富现代韵律的建筑肌理。顶部适度夸大竖向比例的折面女儿墙，宛若坚硬挺拔的龙脊，直冲云端。酒店形象鲜明，尊贵大气，是为整个建筑群落的点睛之笔。

In China, people always bow down to Dragon Totem, so facade design of the Hotel is exactly conceived from the name of "Dragon Town" of Changzhou. Modern design materials and languages are applied in the facade design as metaphor for dragon. The Hotel, office building and SOHO are integrated as a whole through element of "Dragon Scale". Meanwhile, these three buildings respectively take different parts of the dragon as design elements, forming a building complex in the image of "Dragon". The application of glasses in large area on the building facade makes the Hotel become "Head of the Dragon" with the most special image among the three. Glittering and translucent, the Hotel is the dominant unit in the building complex. Bay windows with folding surface spread through the building facade are well organized to build a building texture with modern rhythm. Parapet wall that appropriately exaggerate the vertical proportion on the top directly goes up to the sky, just like the hard and tall keel of the Dragon. The Hotel has distinctive image and magnificent qualities, certainly becoming the punch line of the building complex.

INTERIOR
酒店内装

05

04 酒店入口
05 酒店大堂

06

LANDSCAPE OF HOTEL
酒店景观

酒店后庭院打造成一个自然怡人的花园，休闲的木平台依水而建。在水面远处坐落着一个休憩凉亭，景亭前面是跌水，背面是茂密的树丛，营造一处世外桃源。

Backyard of the Hotel is designed to be a natural and pleasant garden, where timer leisure platform is built by the waters and a pavilion embraced by water drops at the front and dense trees at the back is built on the waters afar, creating a land of idyllic beauty.

07

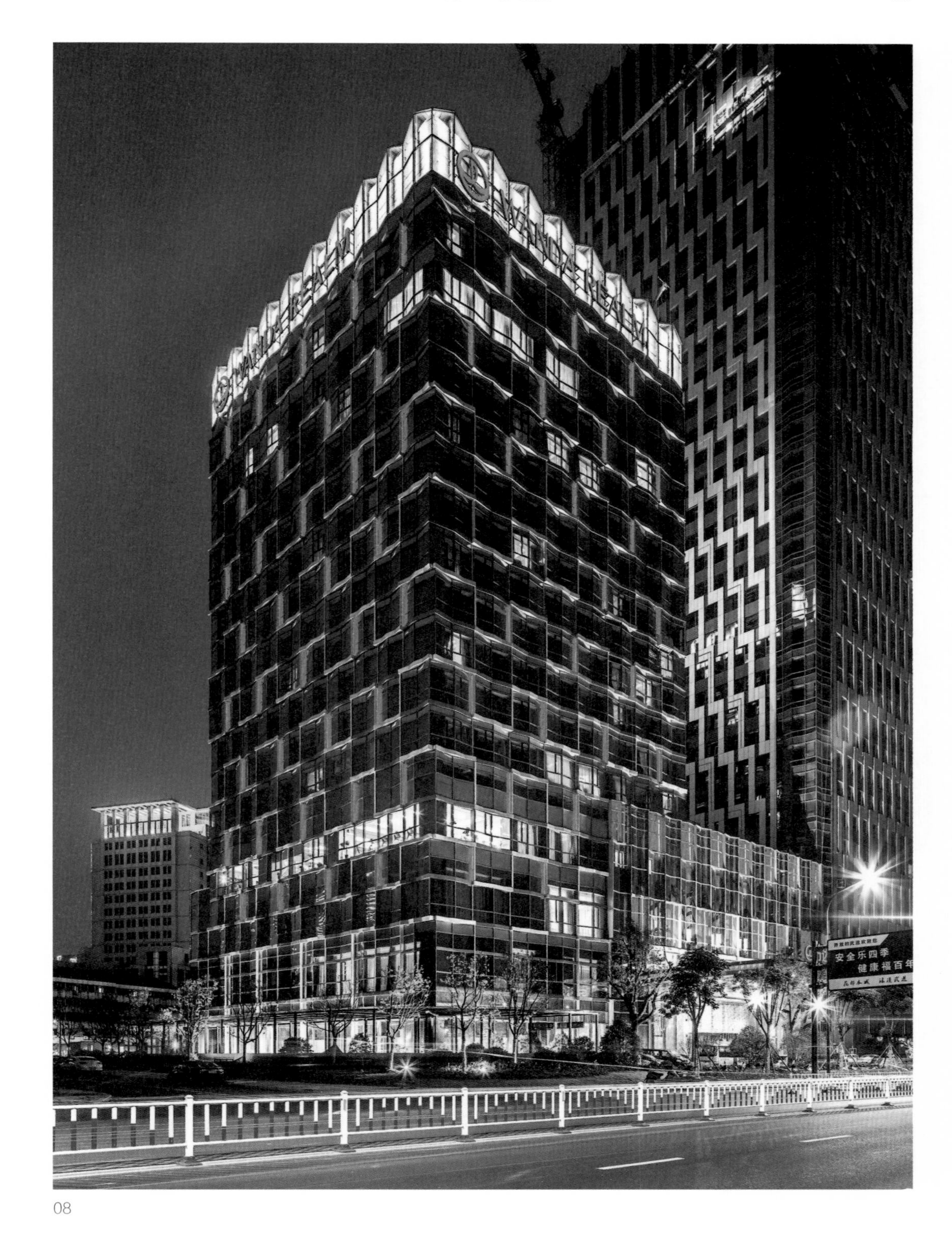

08

NIGHTSCAPE OF HOTEL
酒店夜景

06 景观绿化
07 酒店水景
08 酒店夜景

IMAX

WANDA CINEMA
万达影城

WANDA
COMMERCIAL
PLANNING
2014

WANDA CINEMA
万达影城

万达文化产业集团万达院线副总裁
兼项目建设中心总经理　包文

万达影城在多年经验积累的基础上进行了自我突破创新——万达影城2.0时代。万达影城2.0时代以电影工业为设计主题，结合科技、文化、环保等创新要素，为观众创造出一个全新的观影空间——社交型影城。

在营运模式上，设计伊始就打破以往封闭的经营模式，敞开观影走廊，将原有的"鸡肋空间"转化为可赢利空间。同时，在部分影城设置了自有品牌"ONE"餐厅与"万影随行"衍生品超市，满足顾客的美食及购物需求，同时可以看电影、品美食、买玩具——这一体化的设计使影城逐渐纳入了线下社交功能。

在科技方面，影城运用大量高新技术与互联网技术，如多媒体显示系统、多媒体互动系统、多点触控媒体系统、魔镜系统等，实现了影城的互联网融入与WIFI全覆盖，同时使影城可以更好、更多地与观众互动，打破以往的单向销售模式，极大地增强了与观众的亲和性。

Based on experience accumulated for years, Wanda Cinema develops into the Wanda Cinema 2.0 through self-breakthrough and innovation. Wanda Cinema 2.0, focusing on the design theme of film industry and combining technology, culture, environmental protection and other elements of innovation, builds social-type cinema, a brand new movie space for the audience.

In terms of operation mode, traditional closed operation mode is rejected at the very beginning of design and an open movie corridor is adopted instead, transforming original "unworthy space" into a profitable space. Meanwhile, private brands, such as restaurant "ONE" and derivatives supermarket "Wanda Cinema with You (WAN YING SUI XING)" are set at some cinemas to satisfy customers' demands for food and shopping. Such an integrated design that supports film watching, food tasting and toy shopping at the same time gradually endows the cinema with offline social functions.

01

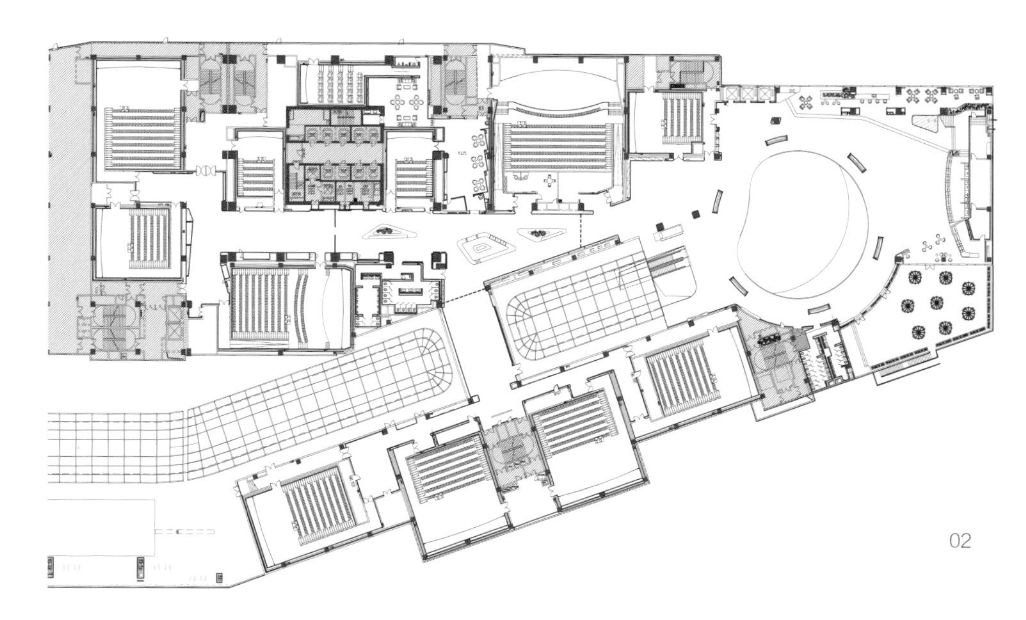

In the aspect of technology, numerous advanced technologies and internet technologies, such as multimedia display system, multimedia interactive system, point touch media system and magic mirror system are applied in Wanda Cinema to realize internet integration and WIFI coverage of cinemas, and better and more frequent interaction between the cinema and customers. The technology as such helps to break traditional unidirectional sales model and greatly enhance affinity of cinemas among the audience.

02

01 通州万达广场影城大厅
02 通州万达广场影城平面图

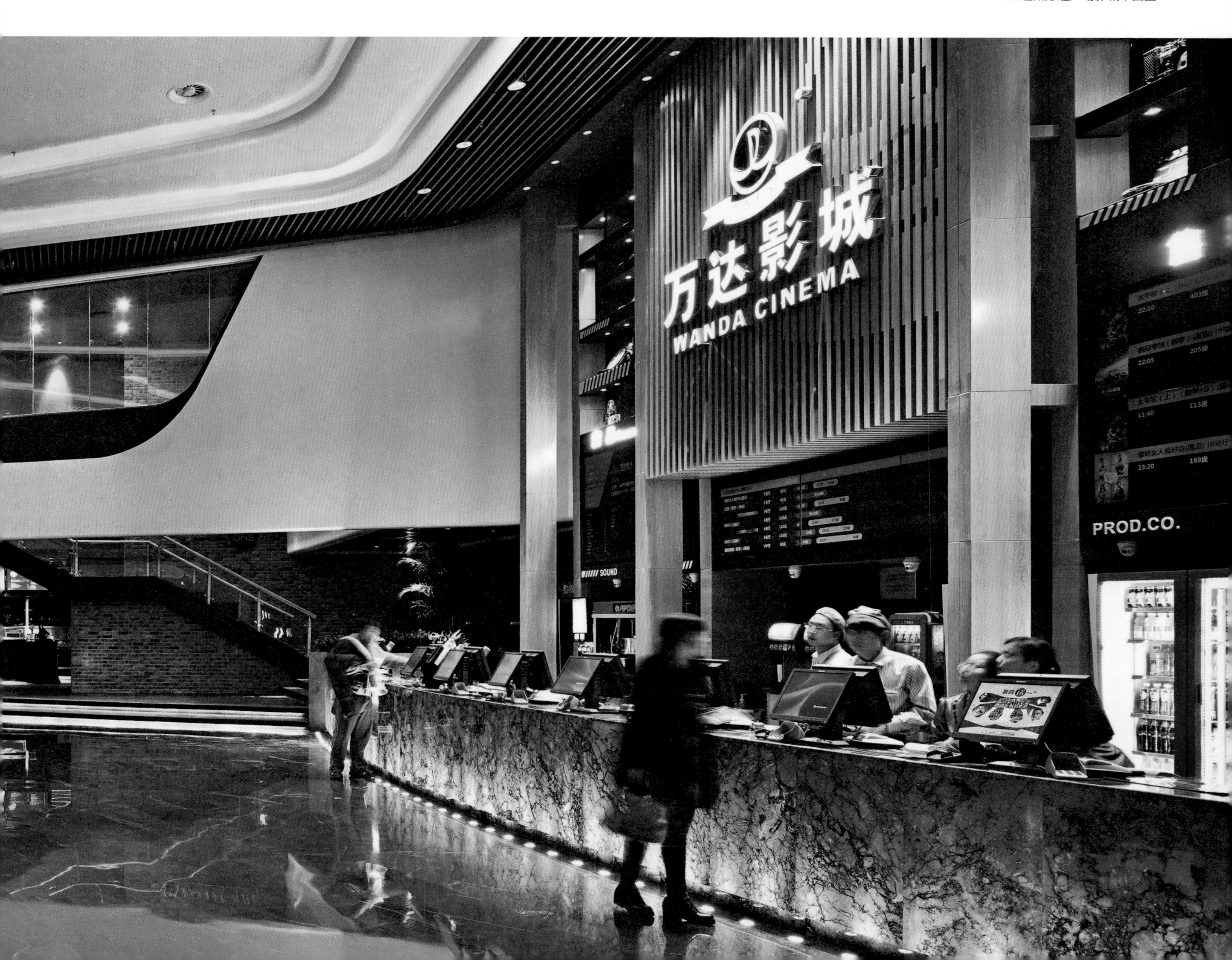

03

04

在文化方面，设计融入了大量的电影艺术元素与电影文化元素、经典的电影名句、经典场景重现、电影发展史科。这些内容普星罗棋布在影城各处，让观众能够漫步电影历史、欣赏经典影片、领略大师风采，做到让电影成为人们生活必备的元素。

在环保方面，万达影城是国首家全方位使用免冲洗无水小便斗，每年节约了大量宝贵的水资源同时减少了碳排放，充分体现了一个国际化企业应有的社会担当，同时也创新地在影城内布置了PM2.5监控系统，充分保证观众的身体健康。

In terms of culture, a great deal of art elements and cultural elements of films, classic film lines and scenes recurrence, and film development historical science are integrated in the design and distributed all over the cinema to allow audience experiencing the film history, watching classic films, appreciating masters' wisdoms, and make film become a life necessary element.

As for environmental protection, Wanda Cinema is the pioneer to use water flushing-free urinal in an all-round manner all over the country, benefiting from which large quantities of valuable water sources are saved and emission of carbon are simultaneously reduced every year, fully undertaking due social obligations and responsibilities of an international enterprise. Moreover, PM2.5 monitoring system is creatively arranged in Wanda Cinema to guarantee health of the audience.

03 通州万达广场影城装饰墙
04 通州万达广场影城走廊
05 通州万达广场影城走廊

George Lucas
THE DREAM BEGAN IN THE SCRIPT AND ENDED WITH THE MOVIE
梦想始于剧本 而终结于电影
1905

万达影城
WANDA CINEMA
IMAX
卖品 Concession Stand
万达
影城
WANDA
CINEMA
万达
WANDA
Box Office

07

06 昆明西山万达广场影城大厅
07 昆明西山万达广场影城走廊
08 昆明西山万达广场影城平面图

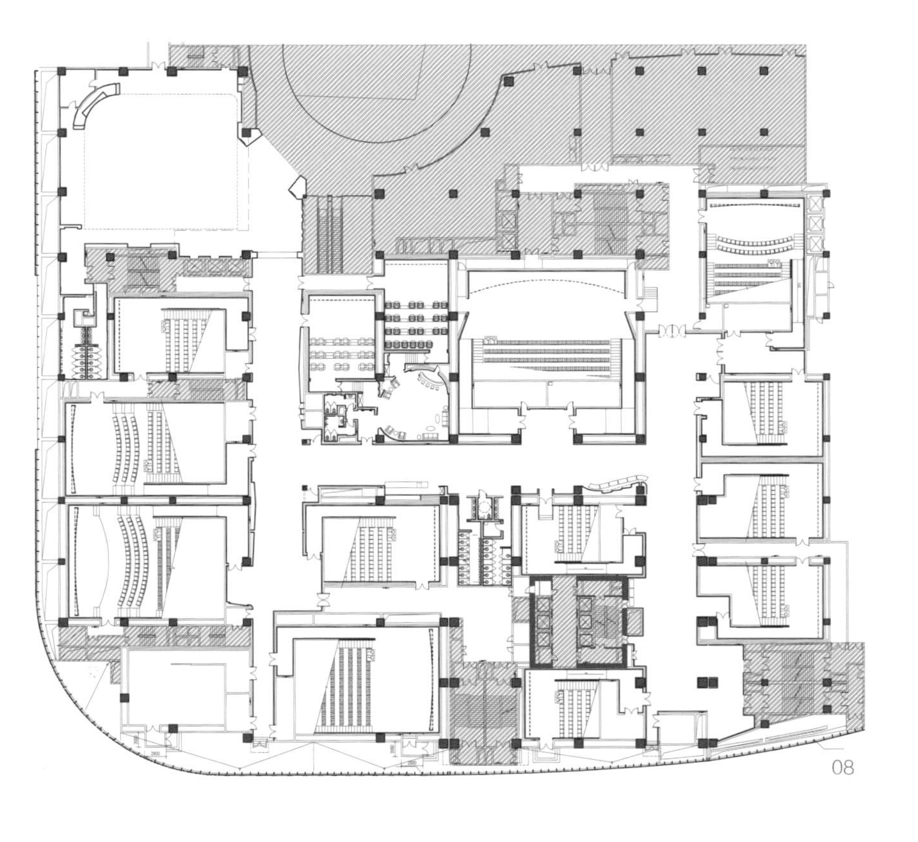

08

09 昆明西山万达广场影城放映厅
10 昆明西山万达广场影城 VIP 放映厅
11 昆明西山万达广场影城 VIP 接待区

09

10

11

万达影城
WANDA CINEMA
IMAX

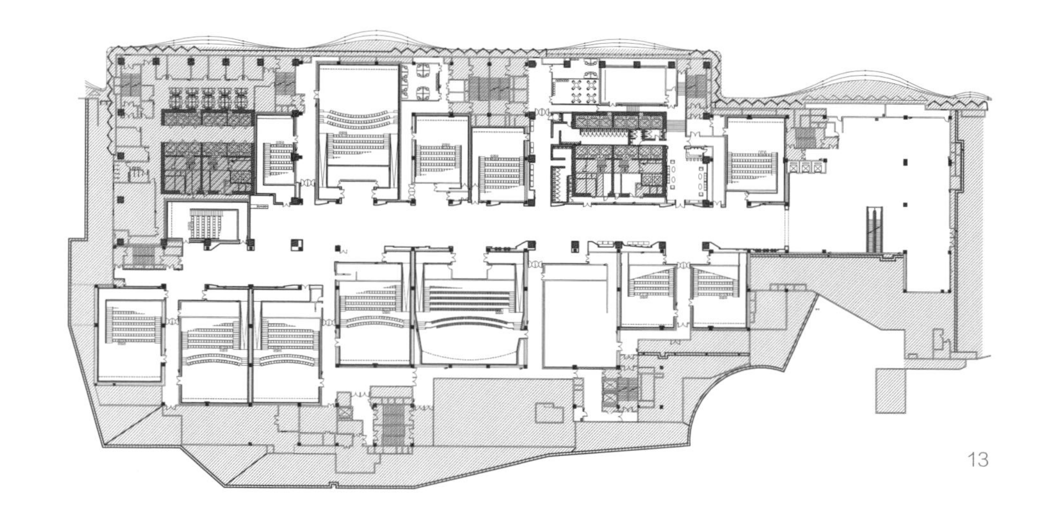

13

12 南宁青秀万达广场影城大厅
13 南宁青秀万达广场影城平面图
14 南宁青秀万达广场影城休息区

14

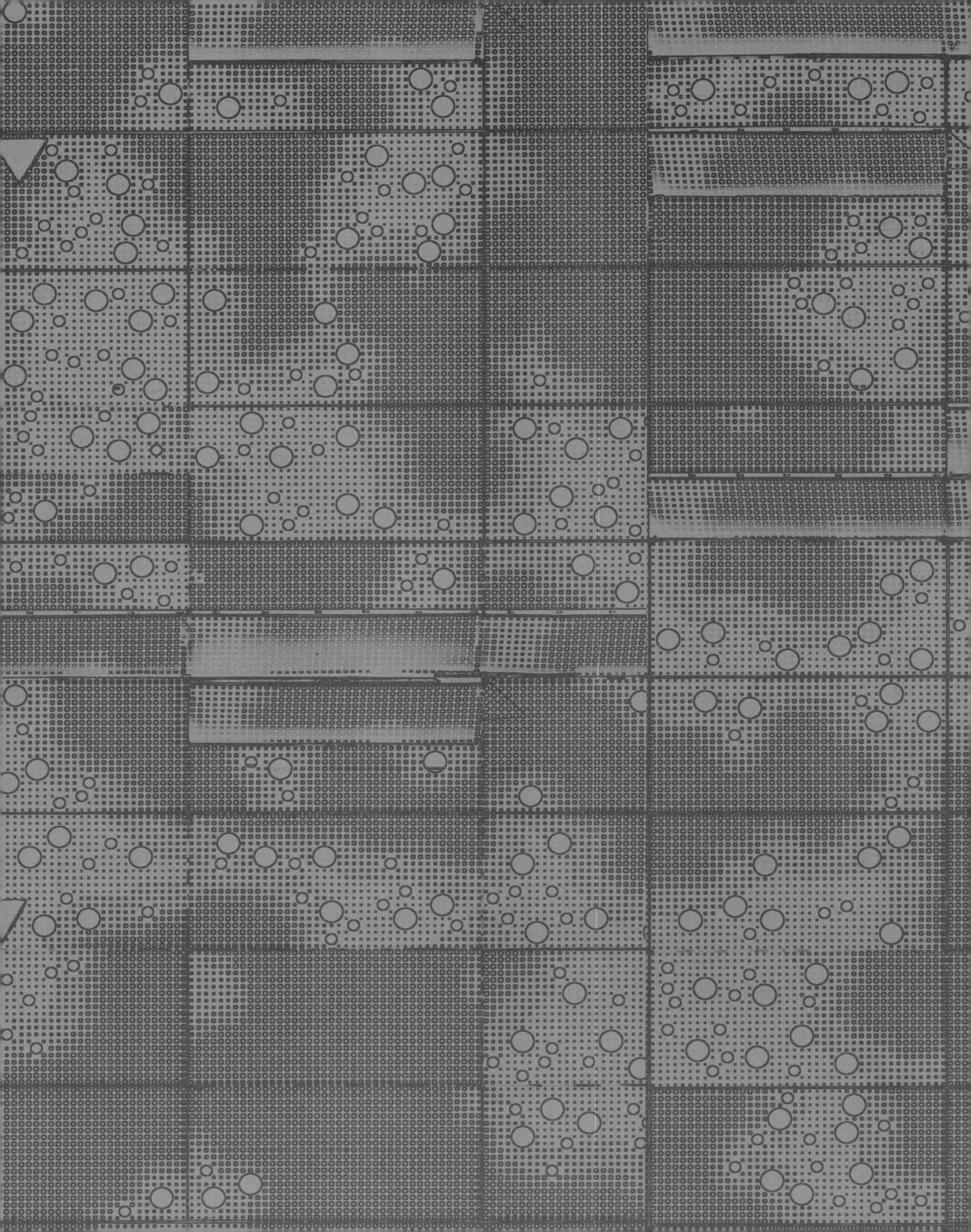

DESIGN AND
CONTROL
设计与管控
WANDA
COMMERCIAL
PLANNING
2014

PERFECTION OF COMMERCIAL PROPERTIES' ECOLOGICAL CHAIN UNDER TURNKEY CONTRACT MODEL
总包模式完善商业地产生态链

万达商业规划研究院常务副院长　朱其玮

万达商业地产宣布从2015年开始，所有万达广场综合体新开工项目都将实行“总包交钥匙”管控模式。董事长王健林表示，“‘总包交钥匙’模式是一项革命性的创新之举，不仅有利于推进企业的反腐败工作，而且能实现多赢局面，使万达商业地产建设生态链更加健康，对整个地产行业起到示范和引领作用。”

一、“总包交钥匙”模式

“总包交钥匙”模式，就是万达在建设工程中，只面对总包单位，不再直接面对分包单位。由原先万达对项目工程一管到底转为由承包公司一包到底——包括项目的计划、质量、安全、成本等事宜——最后向万达交钥匙。总包单位在万达合格“供方品牌库”里，自行选定分包单位、材料设备供货商，并执行万达采购数据中的价格，在建设完成后向万达交钥匙。万达取消招标职能，只需关心交付的成果，不用过多介入项目建设过程。欧美等发达国家的主要工程企业均采用了这种模式，但在中国工程建设领域，万达是首家系统采用这种模式的企业。经过20多年发展，万达商业已建立一套成熟的订单式开发与管理模式，形成了完整的商业地产产业链，实现了信息化、模块化的计划管控，形成了门类齐全的“供方品牌库”；同时，万达与“中字头”施工单位建立了长期战略合作伙伴关系，其他都是中国建筑市场上最高水平的队伍。万达已具备实施“总包交钥匙”管控模式的条件（图1）。

设计阶段			总图	方案及深化	初步设计	施工图	工艺深化图纸		样板段	验收
							土建总包类	专项分包类		
部门	万达集团	商业规划院	▲	▲	○	○				
		项目公司			▲	▲	○	○	○	▲
	总承包商	总包设计部					▲	▲	▲	

注释：▲ 主责方　○ 审核方

（图1）持有物业规划设计权责界面

二、设计管理总包模式

“设计管理（总协调）”是指的主体设计单位对项目的所有设计单位的设计过程以及项目的设计结果进行把控和管理，设计管理单位与甲方除签订自身设计合同外同时承担对其他设计单位的管理和协调工作。“设计总包”是指的业主将该项目所有设计合同

It has been announced by Wanda Commercial Properties that from 2015, the "Turnkey Contract" management and control model will be fully implemented to all newly-opened Wanda Plaza complex projects. As Chairman Wang Jianlin says, "the Turnkey Contract model is a revolutionary measure of innovation. Its implementation not only improves the anti-corruption work of Wanda, but also helps to achieve a multi-win situation, making the "ecological chain" of Wanda Commercial Properties construction much healthier and setting an example for and lead the whole industry."

I. TURNKEY CONTRACT MODEL

Under the Turnkey Contract model, in construction projects, Wanda will only need to face the main contractors and does not need to talk to subcontractors directly any more. Originally, Wanda takes charge of every aspect of a whole project. Now, the contractor shall take charge of the whole project, including issues like project design, quality, safety, costs, etc, and provide a turn-key service to Wanda. The main contractor shall select subcontractors, material and equipment provided from Wanda's "Vendor List", adopt the prize marked in Wanda's purchasing database, and upon completion of project provide a turn-key solution to Wanda. Wanda shall be free from bidding and only care about the delivered results, without much intervention into the project construction process. In the developed countries like US and Europe, most major construction enterprises have adopted this model. However, in the domain of China's engineering construction, Wanda is the industry pioneer for adopting this model. After 20 years' development, Wanda has set a mature order-oriented development and management model, formed a complete commercial properties industrial chain, achieved informationized and modularized planning control and established a complete range of "Vendor List"; in the meanwhile, Wanda has established long-term strategy cooperation partnership with state-owned construction corporations and other construction companies of the highest level in Chinese construction market. It's reasonable to conclude that Wanda has already been geared up for implementing Turn-key Contract model (see Fig.1).

II. MAIN CONTRACT MODEL OF DESIGN MANAGEMENT

Design Management (Main Coordinator) model refers to the model under which the main design institute controls and manages the whole design process of all design institutes and the design results of the project. The main design management institute, in spite of signing its own design contract with the Client, shall also undertake the responsibility to manage and coordinate the work of other design institutes. Design Management Main Contract model refers to the model under which the proprietor sign the whole design contract to the main design contractor and the main contractor shall be responsible for the design results. The biggest difference between the two models lies in the

交给承接单位，由承接单位负责设计结果。两者最大的区别在于“设计管理（总协调）”的乙方公司不承担其他公司的筛选和商务谈判，在设计的管理和协调方面和设计总包一致。

目前国内除了政府大型项目由于招标手续的繁琐采用设计总包的方式外，一般都采用设计管理的方式。万达商业地产的设计总包也是采取这种方式。通过培养几家设计总包单位（比如土建设计单位作为总协调单位），统领各专项设计单位（外立面、幕墙、内装、景观、夜景照明、导向标识及弱电智能化等）。施工图设计单位（总包）负责对各专项设计成果审核并对相关内容图纸“双签”出图，同时设计人应在施工图设计中为此部分内容预留、预埋及配套设计，并协调、配合各单项设计单位完成施工图修改。设计总包单位参与各主力店、次主力店、步行街商铺的内装施工图评审；大商业土建设计单位应对乙装及丙装设计单位图纸进行审核，并负责签字出图，蓝图应加盖大商业土建单位审核章及相关专业注册执业章。设计人应为其他二次设计（包括但不限于室内精装修设计、外立面幕墙设计、外环境设计、夜景照明设计等）提供技术支持与服务，协助其通过政府有关部门的设计审查。总包方负责全面审核本项目所有建筑物（构筑物）的幕墙施工图、夜景照明施工图、内装施工图、景观施工图、弱电智能化施工图、导向标识施工图及相关设计变更等的内容，必须对上述各专项设计的消防、安全以及适用、经济、合理等方面给出审核意见（图2）。

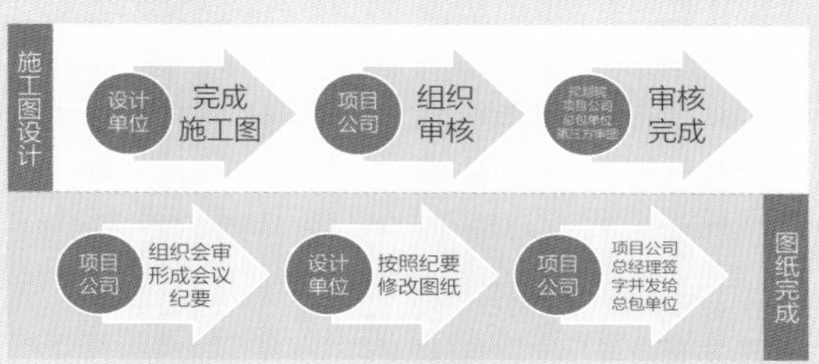

（图2）持有物业设计管控——施工图设计

三、总包模式的创新价值

“总包模式”是万达项目管理的一次重大创新，使万达项目管理达到国际水准。该模式是国际工程企业

（图3）南京江宁万达广场效果图

（图4）南京江宁万达广场实景照片

responsibility of selecting other subconsultants and contract negotiation. Under Design Management (Main Coordinator) model, the Main Design Institute is free from the above-mentioned responsibility. As for design management and coordination, it's the same for the Designer no matter under Design Management (Main Coordinator) model or under Design Management (Main Coordinator) model.

At present, main contract model is only adopted in mega government projects given the tediousness of bidding procedures, while most projects adopt design-management model in China. For its design work, Wanda Commercial Properties Co. adopts main contract model. Several main design institutes are selected and trained (for example, a building design institute is chosen as the main coordinator) to lead and manage other specialized design institutes (facade, curtain wall, interior design, landscape, nightscape lighting, guidance signs, low-voltage-electrical intelligent system, etc.). The design institute of construction documentation (the main contractor) shall be responsible for reviewing the design results provided by each specialized design institute and certifying the drawings of relevant items with "deux signatures"; in the meanwhile, the designer of construction drawings shall make reservation, embedment and coordinate design in construction drawings in consideration of the above-mentioned items of design, and coordinate and cooperate with relevant design institutes to complete the amendment of construction drawings. The main contractor of design work shall take parts in the reviewing process of the interior fit-out construction drawings of anchor stores, sub-anchor stores and business stores; the main commercial civil design institute shall review the drawings provided by consultants and sub-consultants and verify them by signing-off, in addition, the blueprint drawings shall also be stamped with civil institute verification seal and relevant professional registration practitioner seal. The designer shall provide technical support and service for other sub-consultants (including but not limited to interior fitting-out design, façade and curtain wall design, external environment design, nightscape lighting design, etc.) and assist them to pass the design reviews conducted by relevant government authorities. The main contractor shall be responsible for comprehensively reviewing the curtain wall drawings, nightscape lighting construction drawings, interior fitting-out drawings, landscape construction drawing, low-voltage-electrical intelligent system construction drawings, guidance signs construction drawings and relevant design changes of all buildings (structures), and shall make comment on the above-mentioned specialized design from the perspective of fire control, safety, applicability, economical efficiency, reasonableness, etc (see Fig.2).

III. THE INNOVATIVE VALUE OF TURNKEY CONTRACT MODEL

Turn-key Contract model is a great innovation for Wanda's project management, upgrades Wanda's project management ability to the international standard. Such model is an advanced mainstream model of project management welcomed by international construction enterprises. Its superiority is in that the main contractor can play an active role and gain more benefit. Besides, the legal relation between the proprietor and the contractor is relatively simple, with the responsibilities and rights of each party has been clearly-established (see Fig.3 and Fig.4).

The successful implementation of Turn-key Contract model has a strategic significance for Wanda commercial development. In the past, during the construction process of

项目管理的一种主流先进模式，其优越性在于总包能发挥主观能动性，获利较多，业主与总包的法律关系也相对简单，责权利各自清晰（图3、图4）。

实现“总包交钥匙”模式对万达商业的发展具有重要战略意义。过去，在万达广场综合体项目建设过程中，万达要面对众多分包与招投标事宜，管理复杂，耗费大量人力。采取“总包交钥匙”模式后，万达只需面对总包，可集中精力专注于商业模式、产品研发、标准制定、市场营销等更具创新性、知识含量更高的业务，提高管理效率，降低管理成本，进一步提升企业核心竞争力；同时，该模式能从制度上杜绝腐败滋生。过去，万达各综合体项目的负责人，需要耗费大量的精力应付招标与分包事宜，这给腐败行为留出了大量空间。采用新模式后，这些负责人不再拥有相关权力，有利于万达推进企业的反腐败工作。“总包交钥匙”模式真正实现多赢，使万达商业生态链更加合理高效。对于“总包商”来说，采用该模式后，可以增加收入，有效降低建设成本，获得更高利润。而且，由于万达将与“总包”单位共享信息平台，开放数据库，相当于“总包”单位无偿获得万达商业地产十几年积累的宝贵知识产权，可大幅提升其能力层次。“总包”单位成长后，又可支撑万达更快速的发展。对于“分包”单位来说，过去需同时面对万达和“总包”单位，现在只需面对“总包”单位，同样可以节省成本、简化流程。

在规划设计方面，万达以前一直随着商业地产的发展逐步细化，各专业均成立各专业部室，分别指导各项目对应的各专项设计单位。这样，设计管理团队的日益复杂化，让万达规划设计系统逐步膨胀，从2011年的百余人扩张到2014年最高时的近700人。市场变化在于：一是随着商业地产的上市，更加趋向扁平化设计管理；二是随着商业规划设计的逐步成熟，以及轻资产的崛起，管理创新与标准化设计条件成熟。于是“设计总包”应运而生，可以起到如下两个效果：第一，节省成本；第二，同时分享万达这么多年的设计管控经验，也有利于整个设计行业的质量提高。以此，万达规划设计系统的瘦身，甚至外包，也逐渐成为趋势。近期上海万科设计“外包”的出现，就是这种趋势的一个案例。

四、结语

随着我们经济发展和全球经济一体化进程的加速，万达商业项目的设计与施工“总包模式”不断创新发展，必将更快推动商业地产行业的高效发展，带动商业地产生态链更趋完善。

Wanda Plaza complex projects, Wanda has to contract with many sub-contractors and deal with bids and tenders, which is complicated to manage and a waste of manpower. After the implementation of Turn-key Contract model, Wanda only have to contact the main contractor, and can concentrate on more creative and knowledge-oriented business, like business modeling, product research, standard setting, marketing, etc., to improve management efficiency, cut down management costs and further improve the Group's core competency; meanwhile, such model can systematically put an end to corruption. In the past, the people in charge of Wanda's complex projects have to spend abundant energy in dealing with bidding and sub-contracting work, which sets apart a vast space for corrupt behaviors. After the implementation of the new model, those people are deprived of such powers, which shall greatly speed up Wanda's anti-corruption progress. After the implementation of the new model, Wanda can increase the income, and gain more profits by effectively cutting downing construction costs. In addition, since Wanda shall share information platform with the main contract institutes and companies, and give them access to Wanda's database, it means the main contractors shall acquire the valuable intellectual properties accumulated in the past decades of Wanda Commercial Properties for free, which shall substantially upgrade their competence. The development of main contractors will in turn support the development of Wanda at a faster speed. For sub-contractors, in the past, they have to contact Wanda and the main contracts at the same time, but now, they only need to contact the main contractors, which shall cut down costs and simplify the work flow.

As for design planning, in the past, Wanda has subdivided the design planning system along with the development of Commercial Properties, and departments and offices for different specialized projects have been established to guide the specialized design work of corresponding design institute. As a result, with the design management team becoming increasingly complicated, Wanda's design planning system also expands accordingly, undergoing an enlargement from around 100 staff in 2011 to the maximum of around 700 staff in 2014. The changes of market mainly lie in the following two aspects: first, along with the listing of Wanda Commercial Properties, the design management tends to be flattened; second, the commercial design planning is becoming increasingly mature; asset-light projects has been increased quickly; and the conditions for management innovation and design standardization have become mature. Therefore, the Design Main Contract model emerges at the right moment. It shall bring forth the following two effects: first, cutting down costs; second, sharing the design control and management expertise accumulated by Wanda in the past years, which shall help improve the quality of the whole design industry. On this account, it has become a tendency for Wanda to downsize its design planning system and to implement outsourcing. Recently, Shanghai Vanke has also started to outsource its design work, which is an embodiment of such tendency.

IV. CONCLUSION

With Wanda's increasing speed of economic development and global economic integration, Wanda's commercial projects main design contract and construction Turn-key Contract model shall undergo continuous innovation and development, which shall definitely accelerate the development of commercial properties industry and bring along a further improvement of the commercial properties ecological chain.

COMPILATION OF CONSTRUCTION STANDARD 2015 (EDITION A, B, C & D) 2015版建造标准(A、B、C、D版)编制

万达商业规划研究院副院长 侯卫华

2014年底集团完成《2015版建造标准(简约版)》,但随着集团轻资产模式的发展和国家最新《建筑设计防火规范》在2015年5月1日正式实施,新"建造标准"的修订迫在眉睫。在集团领导的统一部署下,由规划院、集团成本部牵头集团其他各相关部门,在《2013版建造标准》基础上,共同修编完成了《2015版建造标准》。

回顾万达商业地产发展过程中,为了处理好如何控制好成本,以及让支出在合理的范围内,支持项目开发中工程、设计以及销售指标的实现,万达建立了一个完善的成本管理体系。这个体系,通过系统性、创新性的成本管理,对企业支出进行严格控制,保证公司经营目标的实现。根据成本划分类型,目标成本分为工程成本和非工程成本。其中工程成本包括土地费用、政府行政事业性收费和建造成本。非工程成本包括各种费用、税金和其他成本。其中建造成本占目标成本的比例达到80%以上,所以如何控制建造成本成为项目成败的关键。

万达的建造成本管理是紧紧围绕"建造标准"而展开,先按照万达"建造标准",由多部门共同制定合理可行的目标成本;再在项目开发过程中通过合约规划对成本进行层层分解、落实;同时依据各项制度、过程管控措施保证不超目标成本;最终保证企业预期利润的实现,达到主动和事前控制的目的。所以,制定准确的"建造标准"尤为重要。

本次"建造标准"修编工作有以下几个主要特点。

1 标准简化原则

《2013版建造标准》原有重资产A版、B版、C版三个版本,加上新增加的轻资产版、轻资产3C版、轻资产D版,将会有6个版本的"标准"同时实行。不同标准之间的级差较小,各部门执行起来相对复杂;同时根据集团发展定位,今后将主要发展轻资产项目,重资产项目将逐渐减少。因此,在集团领导的指示下,将原建造标准中的轻、重资产标准进行合并形成A版、B版、C版、D版四个版本。四个版本的建造成本成阶梯递减,可适应集团对于各级城市项目投资定位的要求,满足各级城市项目投资回报率的需要(图1)。

Although Wanda Group has completed the compilation of *Construction Cost Standard 2015 (Simplified Edition)* by the end of 2014, with the development of the Group's asset-light model and the latest national *Code for Fire Protection Design of Buildings* to be officially implemented on 1st May, 2015, it has become extremely urgent to compile a new construction cost standard. Under the unified arrangement of the Group leaders, the compilation of *Construction Cost Standard 2015*, led by the Planning Institute and the Group's Cost-control Department and with the cooperation of other relevant departments, has been completed based on *Construction Standard 2013*.

Retrospect to the development history of Wanda Commercial Properties, in order to control the costs and to support the realization of construction, design and sales indexes during the project development process with the expenditure being kept within the reasonable range, Wanda has established a sophisticated cost control and management system. The system, through systematic and creative costs management, manages to strictly control the expenditure of the Group and ensures the realization of the Group's operation target. According to different types of costs, target costs are divided into engineering costs and non-engineering costs. Wherein, engineering costs include land costs, government administrative and institutional expenditure and construction costs, while non-engineering costs include all kinds of expenses, taxes and other costs. Normally, construction costs amount to over 80% of the whole target costs; therefore, the key to the success of a project lies in the control of construction costs.

The construction costs management of Wanda focus closely on the Construction Cost Standard. Reasonable and workable target costs shall firstly be made by several departments in accordance with Wanda's Construction Cost Standard; then, in the process of project development, the costs will be broken down and put into practice step by step based on contract planning; meanwhile, it shall be ensured that the actual expenditures shall not exceed the target costs through the implementation of different regulations and process management and control measures; at last, the expected profit shall be achieved, which shall satisfy the demand of initiative and feed-forward control. Therefore, it's especially important to formulate an accurate Construction Cost Standard.

The compilation of the said Construction Cost Standard has several main characteristics, listed as follows:

1. PRINCIPLE OF STANDARDIZATION AND SIMPLIFICATION

As for *Construction Cost Standard 2013*, there are three different editions for asset-heavy projects, namely Edition A, Edition B and Edition C, and plus the newly added Asset-light Edition, Asset-light Edition 3C and Asset-light Edition D, there will be 6 editions of "standard" implementing at the same time. Since the differential between different standards is not evident, it's relatively complicated for different departments to execute; meanwhile, given the development orientation of the Group, the focus of future development will be put on asset-

业态	专项		单位	合并侯建造标准			
				A版	B版	C版	D版
购物中心建造标准	外立面	裙房	元／M^2	1200	880	700	600
	步行街内装		元／M^2	2350	2100	1590	1240
	玻璃采光顶		元／M^2	3600	3230	3230	3230
	夜景照明（建筑及景观）		元／M^2	60	240万	200万	150万
	园林景观	道路及广场	元／M^2	400	350	280	260
		绿化	元／M^2	350	260	230	200
		室外设施及小品	元／M^2	200	100	100	90
		红线外景观	元／M^2	450	300	300	270
	屋顶花园（含美化）		元／M^2	200	80	60	0
设计部	商业部分	外立面方案至施工图	元／M^2	40	30	25	25
		夜景照明	元／M^2	160	60	60	60
		夜景照明动漫	元／M^2	60	/	/	/
		景观	元／M^2	55	40	40	40
		步行街内装	元／M^2	150	130	130	130

（图1）合并后的建造标准

业态	专项	分项	单位	合并后建造成本			
				A版（9+4）	B版（9+4）	C版（7+0.6）	D版（7+0.6）
购物中心建造标准	效果类	单价	元／M^2	1063	808	908	761
		总价	万元	9567	7273	6362	5331
		占比	%	/	12%	16%	14%
		降幅	%	/	24%	13%	16%
	机电（不含小市政及红线外机电费用）	单价	元／M^2	980	970	1159	1157
		总价	万元	12743	12613	8809	8793
		占比	%	/	20%	23%	24%
		降幅	%	/	1%	30%	0.2%
	结构（不含地下四大块，按6度设防，不做人防考虑）	单价	元／M^2	地上 1100 地下 2100	地上 1100 地下 2100	地上 1130 地下 2300	地上 1105 地下 2300
		总价	万元	18300	18300	9290	9110
		占比	%	/	29%	29%	25%
		降幅	%	/	0	49%	2%

（图2）A/B/C/D版建造标准造价对比

2 结合集团成本优化、三年清盘成果和轻资产模式定位

由于万达广场经过多年发展，商业模式日趋成熟，设计风格更趋理性，所以本次修编工作对效果类成本进行了较大幅度的缩减；在保证功能需要的前提下对于立面、景观、内装、夜景专业的非重点部位进行了较大幅度的成本优化（图2）。

3 结合设计标准化、模块化、产业化原则

为全方位优化设计施工成本，真正通过设计创造效益，本次修编工作结合设计科研，从标准化、模块化、产业化方面对项目进行研究。经过集团及规划院领导的多轮指导，指明优化方向、确定优化原则，以设计标准化为源头，进而推进部品模块化，最终达到建造施工产业化；减少因非标准化而引起的成本增加情况，降低项目建造成本；为集团今后大规模迅速推进"轻资产"项目打下了坚实的基础（图3、图4）。

（图3）符合建造标准的轻资产标准外立面

light projects, and the number of asset-heavy project will be reduced. Therefore, under the guidance of group leaders, the original construction cost standards for asset-light and asset-heavy are combined, and compiled into four editions, namely Edition A, Edition B, Edition C and Edition D. These four editions, with laddered-down applicable construction costs, shall satisfy the requirements of different levels of city project investment positioning and the demands of ROI for different tiers of city projects (see Fig.1).

2. COMBINING GROUP COSTS OPTIMIZATION, THREE-YEAR ROUND UP ACHIEVEMENTS AND ASSET-LIGHT MODEL POSITIONING

After years of development, Wanda has established an increasingly mature business model, and the design style has become more rational, therefore, in this revision, costs intended for visual effects have been greatly cut down; under the premise that functional requirements are not compromised, the costs for less-importment components of façade, landscape, interior decoration and nightscape have been drastically optimized (see Fig.2).

3. COMBINING PRINCIPLE OF DESIGN STANDARDIZATION, MODULARIZATION AND INDUSTRIALIZATION

To optimize the costs of design construction at all dimensions and to truly achieve benefits through design, design research has been integrated into this compilation work, and the project has been studies from the perspective of standardization, modularization and industrialization. Under several rounds of guidance from group leaders and leaders with the Planning institute, the orientation for optimization has been specified and the principle for optimization has been determined; the optimization shall start from the standardization of design, which shall advance the modularization of components, and finally help achieve the industrialization of construction

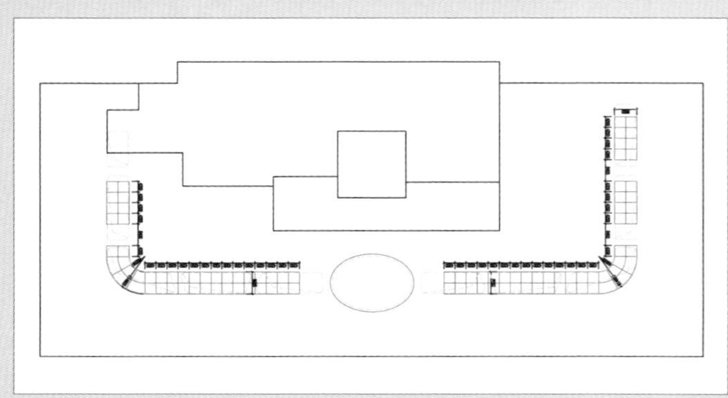

（图4）采光顶通过设计手段控制在建造标准内

事项	专业	关键词	新旧规范变化	增加成本	最终增加
消防原因增加成本项	建筑	玻璃采光顶	开启扇面积由 10% 增加到 25%	240万	－240万（设计优化）
		消防电梯	每个防火分区至少设置一部消防电梯	36万	0
	内装	室内步行街两侧商铺玻璃	由钢化玻璃调整为防火玻璃	80万	－150
	设备	风管防火保护措施	防火阀 2 米范围内采用耐火风管或采用防火保护措施	28万	－7.5（慧云优化）
	电气	防火门控制系统	增加疏散通道防火门控制系统	50万	
		消防配电线路	消防配电线路与其他配电线路分开敷设在不同的电缆井、沟内、或采用矿物绝缘电缆	26万	
		消防供电	消防供电引自同一个区域变电站时，应设柴油发电机组（建议所有项目统一考虑）	290万	－27.5
外立面				0	－30
景观				0	0
成本增量				750万	295万

（图5）A/B/C/D版建造标准造价对比

4 结合集团各级领导的工作指示和新标准新规范的内容

本次修编还将集团各级领导在以往项目实施过程中涉及“建造标准”的重要指示进行增补。对于实施的新规范、新标准等方面，涉及“建造标准”方面变化的事宜也进行了修编（特别是对于新版《建筑设计防火规范》等涉及消防安全规范的事宜进行了重点修编），为使新编“建造标准”既满足集团对于建造成本方面的控制要求，又能满足新“规范”对于消防方面的要求进行了多次讨论，最终取得了满意的结果（图5）。

5 结合集团各业态、各部门对于建造标准方面的需求

本次修编充分考虑集团各业态、各部门对于建造标准方面的需求，本着为保证项目顺利运营的原则，在保证建造成本可控的情况下，尽量考虑各部门的运营需求，对于旧版“建造标准”中不符合运营要求的内容进行了修订。

《2015版建造标准》修编工作历时半年，在集团成本部和各相关部门的大力支持下完成，必将为集团各部门的顺利开展设计、工程建设工作提供成本依据，为总体控制各项成本保驾护航。

and building; the costs increase caused by non-standardization shall be reduced in order to cut down the construction costs of project; a solid foundation shall be laid for the Group's large-scale and rapid advancement of asset-light projects in the future (see Fig.3 and Fig.4).

4. COMBINING INSTRUCTIONS FROM GROUP LEADERS AND THE CONTENTS IN NEW STANDARDS AND SPECIFICATIONS

In addition, key instructions regarding the construction cost standards given by Group leaders at different levels during the previous construction process have been added into this revision. As for newly-implemented specification and standards, matters concerning construction cost standards have also been edited and revised (special and careful attention has been given to the revision of matters concerning fire safety specifications mentioned in the new edition of *Code for Fire Protection Design of Buildings*). To satisfy both the Group's requirements on construction costs control and the fire protection requirements of the new cord, the newly-compiled Construction Cost Standard has been repeatedly discussed, and finally achieving a satisfactory result (see Fig.5).

5. COMBINING THE REQUIREMENTS ON CONSTRUCTION STANDARDS FROM THE GROUP'S DIFFERENT TYPES OF BUSINESS AND DIFFERENT DEPARTMENTS

This revision has fully considered the requirements on construction standards from the Group's different types of business and different departments. Based on the principle of ensuring project's successful operation, under the premise of controlling construction costs, operational requirements of different departments have been taken into consideration, and contents in the previous Construction Standard that are at variance with operational requirements have been revised.

It takes half a year to finish the compilation of *Construction Cost Standard 2015* with great support from the Group's Cost-control Department and other relevant departments. It's believed that the new standard will certainly provide a cost basis for carrying out design and project construction work smoothly for different departments of the Group, and escort the overall control of each item of costs.

INITIAL PLANNING AND COORDINATION OF COMMERCIAL PROPERTIES
商业地产前期规划及对接

万达商业规划研究院副院长　孙培宇

万达商业地产前期规划及对接，是指规划院与集团前期发展相关部门的对接。2015年，是万达商业地产的转型年，是由“重资产”向“轻资产”转型的一年，而前期发展业务的参与部门也相应发生了变化，具体的前期对接内容和工作方式也发生了一些变化。下面先从之前“重资产”模式的前期发展对接开始介绍。

一、重资产模式的前期发展对接

“重资产”模式是以销售物业提供现金流支持持有物业（商业、酒店）建设的模式。对于万达来说，每一个开发项目不是要完成多少项目本身的盈利，而是要通过房地产销售将自持物业投入的成本平衡。所以一个万达“重资产”项目在前期是否成立的一个重要因素就是项目总成本和总销售额是否可以打平或者接近平衡。

在这样的前提下，“重资产”的前期参与部门主要是发展中心、规划院、成本控制部；发展中心负责项目的选址、政府谈判；规划院负责选址的考察、总图设计和指标；成本控制部负责根据发展中心提供的政府地价、规划院提供的总图指标以及对项目周边的房产市场价格进行项目现金流测算。所以规划院在前期对接的主要部门就是发展中心和成本控制部（图1）。其中，对发展中心的对接分为技术文件对接和配合发展中心进行政府谈判两部分。

技术文件对接根据项目进行的阶段又包括了项目交底（补充交底）、总图指标的移交、协议性文件的技术类审核和挂牌、摘牌文件的技术类审核。

The initial planning and coordination of Wanda Commercial Properties refers to the coordination made by Wanda Planning Institute with relevant departments of the Group's concerning project initial development. The year 2015 will be a transition year for Wanda Commercial Properties to transform from the asset-heavy model to the asset-light model, and corresponding changes will happen regarding the departments participating in initial development and some changes regarding the specific initial coordination contents and protocols. The following instruction will start from the initial development coordination under the previous asset-heavy model.

I. INITIAL DEVELOPMENT AND COORDINATION UNDER ASSET-HEAVY MODEL

Asset-heavy model refers to the model that properties for holding (commercial building and hotel) are funded by the cash flow provided by the sales of properties. For Wanda, what matters for developing a project is not to achieve a fair amount of benefit from the projects themselves, but to achieve a balance for the costs of self-owned properties investment from the sales of properties. Therefore, when deciding whether to establish an asset-heavy project in the initial period, one important factor is whether the total sales can equal to or approach the balance of total project costs.

Under such premise, the major departments participating in the initial stage of asset-heavy projects are the Wanda Development Center, the Planning Institute and the Cost-control Department; the Development Center is responsible for project site selection and the negotiation with the government requirements; the Planning Institute is responsible for site test-fit feasibility study, master plan design and index setting; the Cost-control Department is responsible for project cash flow projection, which is based on the government-set land price provided by the Development Center, the master plan and indexes provided by the Planning Institute and the market bench-mark of prosperities around each project. Therefore, in initial stage, the Planning Institute shall mainly coordinate with the Development Center and the Cost-control Department (see Fig.1). Wherein, the coordination with the Development Center mainly concentrates on two parts: technical

（图1）项目前期选址考察

在发展中心对项目进行立项并报董事长通过后，发展中心通过项目交底单的形式向规划院进行项目规划条件的技术性交底。交底内容包括：项目的城市地图、Google地图、项目红线图及现状地形图、该项目有地的控规指标（包括用地面积、控制高度、容积率、绿化率、建筑密度、机动车和非机动车的规划要求、公建占比等）以及人防、日照等相关地方规定；同时尽可能提供用地周边的现状地形图、政府规划、地勘资料等。规划院根据交底内容以及董事长通过立项时对该项目的指标要求进行总图设计，之后报董事长审图会审批总图方案。在总图方案和指标通过之后，规划院经内部会签后向成本控制部通过OA进行总图方案和指标的移交，而成本部则进行该项目的现金流测算，并将测算结果报董事长审图会。如果测算结果达到董事长的要求，规划院内部会签后向发展部通过OA进行总图方案和指标的移交；如没有达到要求，则进行方案指标的调整，再次进行测算；直至符合董事长对项目整体现金流的要求，再向发展中心进行方案移交。对于多次调整仍无法满足测算的项目，即停止该项目或由发展中心与政府沟通地价和对测算影响较大的相关条件。

发展中心拿到规划院移交的总图方案后，由发展中心进行与政府的方案对接和谈判，其中遇到与控规条件冲突或需要向政府领导进行技术类汇报的情况下，规划院会派经验较丰富的人员协助参与发展中心与政府的对接和谈判。对接和谈判的目标，是促使政府接受万达的规划方案和指标。万达项目的规划，因为大型商业的特性，一般会出现建筑密度过大（特别是商业地块需单独出让的情况）、绿化率较小、停车指标较难满足等情况；另外还会出现调整规划路网、与城市控制性规划出现矛盾的情况。规划院要通过向政府的汇报、沟通使政府理解万达项目规划的理由，接受万达的规划方案；同时也要把政府坚持的以及新提出的对该项目的要求和规划条件，反馈回集团以便对规划方案做出调整并上报董事长。这其中，某些项目的汇报和沟通需要进行多个轮次才能够达到双方的认可。

在项目的所有商务和技术内容都获得万达、政府双方的认可之后，项目进入合同、挂牌拿地阶段。在这个阶段，由发展中心将与政府商定的合同、挂摘牌文件等协议性文件通过OA发给集团相关部门（包括规划院、成本控制部、法律事务部、项目中心、商管中心等）进行审核，其中规划院负责对其中的技术部分进行审核并反馈调整意见。

以上是“重资产”模式前期对接的过程、参与部门和各部门职责。

documents delivery and cooperation with Development Center to support their government negotiation.

Technical documents delivery, in accordance with different stages of project, shall include project technical clarification (and supplement technical clarification), delivery of general drawing indexes, technical review of negotiation documents and technical review of land listing and acquisition documents.

A project established by the Development Center shall firstly to submit to the Chairman for review and approval. Upon approval, the Development Center shall carry out technical explanation mainly concerning the conditions of project planning in the form of clarification lists to the Planning Institute. The contents in explanation lists shall include city map of the project, Google map, property-line map and current topography of the project, regulatory requirements of project site (including site area, controlled height, plot ratio, greening rate, density of building, requirements on motor vehicles and non-motor vehicles, public architecture ratio, etc) as well as other relevant local stipulations on civil air defense, sunlight, etc.; meanwhile, the Development Center should provide the current topography of surrounding land, government regional plan, geological research data, etc. The Planning Institute shall carry out master plan design in accordance with the clarification lists and the index requirements instructed by the Chairman when approving the project entitlement. Upon completion, the master plan shall be submitted to the Chairman's drawing committee for review and approval. Once approved, the master plan and indexes, after internal due-check and signing by the Planning Institute, it shall be delivered to the Cost-control Department through OA. After that, project cash flow calculation shall be carried out by the Cost-control Department, and the results of calculation shall be submitted to the Chairman's drawing review committee. If the economic projection results reach up to the requirements of the Chairman, the general drawing plan and indexes, after internal validation and signing by the Planning Institute shall be delivered to the Development Center through OA; if the results fail to reach the requirements, the project indexes shall be adjusted and recalculated to the satisfaction of Chairman's requirements on the overall cash flow of the project, and the project shall be delivered to the Development Center again. If after several attempts of adjustment, the project still fails to satisfy the economic requirement, the project shall be ceased immediately or the Development Center shall negotiate with the government about land price and other relevant requirements that have major influence on economic projection.

After the acceptance of the master plan delivered by the Planning institute, the Development Department shall be responsible for coordinating and negotiating with the government. Should any conflict to the regulation happens or a technical presentation is required to be delivered to heads of government, staff with expertise will be sent by the Planning Institute to assist the Development Department with the coordination and negotiation process. The object of such negotiation is to persuade the government to accept Wanda's plan and indexes. Since Wanda's projects are mostly large scale commercial projects, the problems, including the density of building being too big (especially in the situation that commercial land need to be transferred alone), the greening rate being too small and the parking index being too hard to satisfy, are quite common; in addition, the conflict between regulatory road network planning and the city controlled planning is also likely to happen. After the presentation delivered for the government, the Planning Institute shall try to make the government understand the

二、轻资产模式下的前期发展对接

“轻资产”模式是万达通过融资或他方投资建设的模式建设万达广场，万达不再作为商业项目开发的投资方，商业项目不需要销售类物业提供现金流的支持。在这种模式下，前期发展的工作发生了一些变化，变化主要体现在以下几个方面。

1. 对项目用地要求的变化

“重资产”项目包含销售类物业，用地规模一般在150~200亩（10~13.5公顷）左右；“轻资产”项目用地规模一般在70~80亩（4.6~5.3公顷）左右；同时要求用地直接临城市主路，用地形状要求比较规整。

2. 注重投资回报率

“轻资产”项目的测算主要围绕投资回报率进行测算，对前期方案图纸的深度要求更深、更准确，而且必须要求用单体方案进行最终测算；不仅需要方案指标达到要求，方案的合理性和可实施性也要达到要求。

3. 主责部门改变

测算内容的改变导致测算的主责部门发生改变，投资回报率的测算主要是资产管理部完成。

4. 强化了模块化和标准化

“轻资产”项目大量采用模块化和标准化，导致发展部与政府相关部门的对接变得相对简单；而“轻资产”项目数量的大幅增加，也导致前期工作要加快进度。规划院配合参与政府的汇报和技术对接数量大幅减少，大部分项目发展中心可以自己完成（图2）。

5. 对规划院的新要求

为了配合发展中心前期工作快速、顺利地进行，规划院向发展中心派遣了发展规划小组。小组在项目立项之后马上进行简单的总图、指标设计，并提供给资

reasons for Wanda's project planning and to accept Wanda's plan through communication; in the meanwhile, for the requirements and regulatory conditions that the government strongly insists or brings forth for the first time, the Planning Institute shall report them back to the Group leaders so as to make corresponding adjustments on the plan and submit to the Chairman. Sometimes, some projects thereof shall undergo several rounds of presentations and negotiations before finally accepted by both parties.

After all the commercial and technical contents of the project are approved by both Wanda and the government, the project shall enter the contract-making, land listing and land-acquiring stage. In this stage, the Development Department shall settle down the contract and negotiation documents including land listing and land-acquiring documents with the government. The above materials shall be sent to relevant departments (including the Planning institute, Cost-control Department, Legal Department, Project Management Center, Commercial Management Center, etc) through OA for review and approval. Wherein, upon receiving the materials, the Planning Institute shall be responsible for checking and giving adjustment suggestions regarding the technical perspectives.

The above explains the process, participating departments and responsibilities of each department in the initial stage under the asset-heavy model.

II. INITIAL DEVELOPMENT AND COORDINATION UNDER THE ASSET-LIGHT MODEL

The asset-light model refers to the model under which the construction of Wanda Plaza is funded by financing and investment from other parties. Hence Wanda will no longer be the investor of commercial project development, and commercial projects will be independent of the cash flow support provided by the sales of properties. Under such model, some changes will happen concerning the development work in the project initial stage, which mainly reflect in the following aspects:

1. CHANGES TO THE REQUIREMENTS OF PROJECT LAND USE

Asset-heavy projects include sales properties, and the scale of land use usually amounts to around 150~200 mu (10~13.5 hectares); the scale of land use of asset-light project usually amounts to 70~80 mu (4.6~5.3 hectares); in the meanwhile,

（图2）轻资产项目效果图

产管理部进行初步测算。对于测算结果靠近或达到目标的项目，再移交给规划院进行下一步工作；测算与目标差距较大的项目，则重新选址或根据规划总图反映的问题有针对性地与政府进行对接。同时，发展规划小组还参与发展中心的项目选址、技术条件对接等工作，减少了集团部门之间的对接环节，大幅提高了前期项目的工作效率，减少项目因选址不合适、条件不全面等因素导致的无效或反复工作，大大提高了前期项目推进的速度。

“轻资产”模式刚刚开始，很多的工作还处于磨合和摸索阶段。前期发展工作会通过集团相关部门的共同推进，我们的工作将更加完善、更加紧密、更加高效！

the land shall be close to city main road and the shape of the land shall be relatively regulated.

2. EMPHASIS ON ROI

The calculation of asset-light projects shall concentrate on ROI, which requires the initial planning drawing to be more thorough and accurate, and it's required that the final calculation shall be carried out based on the architectural floor plans; therefore, not only plan indexes shall meet the requirements, the rationality and feasibility of the floor plans shall also meet the requirements.

3. CHANGES OF MAJOR RESPONSIBLE DEPARTMENT

Due to the change of calculation contents, the major responsible department has also been changed. The calculation of ROI is mainly carried out by the Asset Management Department.

4. REINFORCEMENT OF MODULARIZATION AND STANDARDIZATION

Since asset-light projects are largely modularized and standardized, the coordination with relevant government departments becomes relatively simple; while, with the dramatic increase of the quantity of asset-light projects, the progress of initial work is required to be quickened. Therefore, the Planning Institute has greatly reduced its participation in government presentation and technical coordination, and the Development Center is able to complete most of the project feasibility study by itself (see Fig.2).

5. NEW REQUIREMENTS ON THE PLANNING INSTITUTE

To help ensure the fast and smooth progress of the Development Department's initial work, the Planning Institute has dispatched a Development Planning Team to the Development Department. After a project has been approved, the team shall carry out the floor plans from master plan and indexes immediately, and submit them to the Asset Management Department for initial calculation. If the calculation result of a project is close to or meet the objective of the project, it shall be handed over to the Planning Institute for the next stage; if the calculation result is still quite afar from the project objective, the site shall be re-selected, or negotiation shall be made with the government in accordance with the major problems reflected in the general drawings. Meanwhile, the Development Planning Team has also taken part in other work of the Development Department, like project site selection, technical condition coordination, etc, which helps to cut down cross-department coordination and greatly improves the work efficiency of project in the initial stage; in addition, since invalid and repeated work caused by inappropriateness of site location or incompleteness of conditions has been reduced, the speed of project advancement in the initial stage has been speed up largely.

Since the asset-light model has only been implemented for a short time, it is still at the test and trail stage. The initial development work will be advanced jointly with the cooperation of relevant departments of the Group, and our work will become more improved, and tightly managed.

ONSITE DESIGN CONTROL OF WANDA PLAZA PLANNING INSTITUTE
万达广场规划现场管控

万达商业规划研究院副院长　杨旭

万达广场规划的现场管控是项目整体管理重要组成部分，也是设计成果最终落实的重要步骤。现场管控的最终目的是在设计、施工、成本、整体计划等综合条件下，最大程度保障集团认可设计成果的有效实施。

规划现场管控是一项复杂的综合性管理工作，主要的难点在于如何保证在众多项目实施过程中有高标准的统一依据；如何保证设计成果转化的效率与品质；如何解决现场管控中多单位、多系统、多专业、多工种的组织配合；如何评价现场实施效果以及如何将成功经验转化为更高标准的机制并迅速推广等几项问题。

现以效果类管控为例，阐述万达广场现场管控如何通过管理手段有效解决以上几项现场管控的难题。

一、高标准的管控依据

在万达集团2020年实现500个万达广场的目标的快速发展过程中，平均每年有几十个万达广场陆续开业，如何才能保证如此众多的项目都具有较高的标准？

通过集团层层筛选、审核的有效签批设计文件及设计封样，成为项目建设的统一目标，也是现场管控的重要依据。集团的有效签批文件，主要由集团签批下发的以下文件组成：（1）总图方案；（2）平面方案及方案深化；（3）立面方案及方案深化；（4）内装、导向、景观、夜景等效果类签批方案及设计封样。集团的有效签批文件，从源头上确保了现场实施的高标准。

同时，集团现行的制度、技术标准、强条和要点，现行国家、行业、地方的法规和规范，是现场管控的依据，也是现场管控措施制定、变更认定、成果评估考核的基础。

二、重点内容管控

现场管控的主要工作是确保设计成果转化的效率与品质。对于效果类管控，工作内容的重点是材料封样、样板段的控制。只有确保封样材料的标准和样板段的正确实施，才能保障项目的整体品质和大面积迅速施工；同时在施工过程中，现场管控人员应对施工品质进行定期检查，及时修正施工错误，确保设计正确贯彻。

Onsite design control, which is under the charge of Wanda Commercial Planning & Research Institute, is an important component of the overall management of a project and an important step for the final achievement of the design results. The ultimate objective of onsite design control is to ensure the effective implementation of the design results approved by the Wanda group with a comprehensive consideration of the conditions like design, construction, costs, overall planning, etc.

Onsite design control by Planning Institute is a complicated and comprehensive management task, whose difficulties mainly lie in the following aspects: how to ensure the consistency of a high-standard and unified basis during the implementation process of numerous projects, how to ensure the efficiency and quality of design result realization, how to coordinate the work of different institutes, systems, professions and disciplines during onsite control process, how to evaluate the effects of onsite implementation and how to transform successful experience into a mechanism of higher standard and put it into implementation at a rapid speed.

Now, take effects control as an example, the following paragraphs give an explanation on how to effectively solve the above-mentioned difficulties of onsite control through the application of management tools under the Wanda Plaza Onsite Control System.

I. HIGH-STANDARD DESIGN CONTROL BASIS

To achieve Wanda Group's object of opening 500 Wanda Plazas by 2020, several dozens of Wanda Plaza shall be opened successively every year, and in the process of such a rapid development, how to ensure the high standard and quality of so many projects has become a difficult question.

The effective design documents and sealed design material samples that have been carefully selected and approved by the Group have become the unified objective of project construction and the important basis for onsite design control. The signed and approved documents mainly consist of the following items: (1) master plan drawing; (2) floor plans and details; (3) façade elevations and details; (4) approved and signed plans and sealed material samples of interior fitting-out, guidance sign, landscape, nightscape etc.; These effective signed and approved documents help ensure the high standard of onsite implementation for the source.

Meanwhile, the current system, technical standard, mandated regulations and key points of Wanda Group as well as the current national, industrial and local laws and regulations are also the basis of onsite design control, which serve as the foundation for formulating measures, recognizing alterations and evaluating achievements during onsite control process.

1. 材料封样

集团移交的设计封样，是项目公司现场管控阶段材料封样组织的依据。项目公司设计部组织设计单位在初设、施工图阶段对材料封样进行深化及补充。材料封样具有一定的实施标准：要与设计封样一致；种类要覆盖全部材料；材料封样要多种材料进行比选；材料封样应满足总包招分包的要求（图1）。

2. 施工样板段

样板段既是封样材料现场确认的重要步骤，也是施工工艺和重点区域构造节点现场最终确认的步骤。样板段应该严格按图施工。施工封样以集团确认的材料封样为依据，施工封样根据样板段进行比选，通过在现场条件下的比较确定最优材料，签确后作为材料封样。施工封样应与设计封样保持一致。通过施工样板段，对设计封样材料和节点设计进行现场验证，确保大面积实施效果（图2~图5）。

3. 现场检查

现场检查是规划现场管控重点工作内容的延伸和补充。在材料封样和样板段阶段及后续实施过程中，项目公司应组织设计单位、总包单位对现场施工情况进行定期检查；集团相关部门应对现场情况进行抽查，并要求项目、设计供方定期上报现场检查情况，掌握现场进展情况，确保规划管控内容顺利推进。

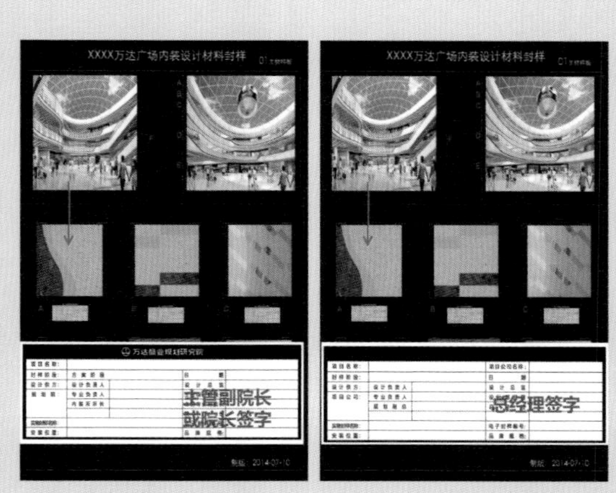

（图1） 设计封样样板

（图2） 样板段区域确定

（图3） 施工封样材料比选

II. CONTROL OF KEY CONTENTS

The major task of onsite design control is to ensure the efficiency and quality of design result transformation. For effects control, the key point is the control of sealed sample material and mock-up construction. Only the sealed sample material and the correct mock-up ensures, the quality of the whole project and the rapid construction at a large scale; meanwhile, during the construction process, the onsite control staff shall check the construction quality regularly to correct construction deficiencies and to ensure the correct implementation of the design in a timely manner.

1. SEALED SAMPLE MATERIALS

The sealed sample materials handed over by the Group shall serve as the basis for the management of sealed sample materials in the onsite control stage of the project company. In initial design and construction drawing design stage, the design institute, under the guidance of the design department of the project company, shall further specify and complement the sealed sample materials. The sealing of sample materials shall comply with the following implementation standard: it shall be in accord with the designed sealed sample; it shall cover all materials; the sealed sample shall be chosen from a compariable variation of materials; the sealed sample materials shall satisfy the requirements that are set for the chosen subcontractors by the main contractor (see Fig.1).

2. MOCK-UP CONSTRUCTION

Mock-up is an important step for verifying the sealed sample materials onsite, as well as the final onsite verification for construction method and important section assemblies. The construction of mock-ups shall be carried out in accordance with the drawings. Based on the sealed sampled materials approved by Wanda Group, after comparing and selecting the mock-up sections, the optimized materials shall be chosen as the final approval and sign-off sample materials for construction based onsite. The sealed sample materials for construction shall be in accord with the designed sealed sample. Through the construction of mock-ups, the design-sealed sample materials and the assembly details shall be verified onsite to ensure the large-area implementation effects (see Fig.2 to Fig.5).

3. ONSITE INSPECTION

Onsite inspection is the extension and supplementation of the key contents of onsite control. In the stage of sample materials sealing, mock-up construction and the process of follow-up implementation, the design institute and the main contractor shall carry out regular inspection on the conditions of onsite construction under the organization of the project company; relevant departments of the Group shall carry out spot check on the onsite conditions, and require the project company and design provider(s) to report the onsite inspection conditions regularly, in order to grasp the onsite construction progress condition and to ensure the smooth advancement of the onsite control contents handled by the Planning Institute.

III. ORGANIZATION AND COORDINATION OF WORK DIVISION

The executing subjects of onsite control mainly consist of the Wanda headquarter, the project company, the design provider(s) and the main contractor, each controlling unit taking charge of different professions and types of work; meanwhile, different systems may implement an intersecting management model for the same control content. Therefore,

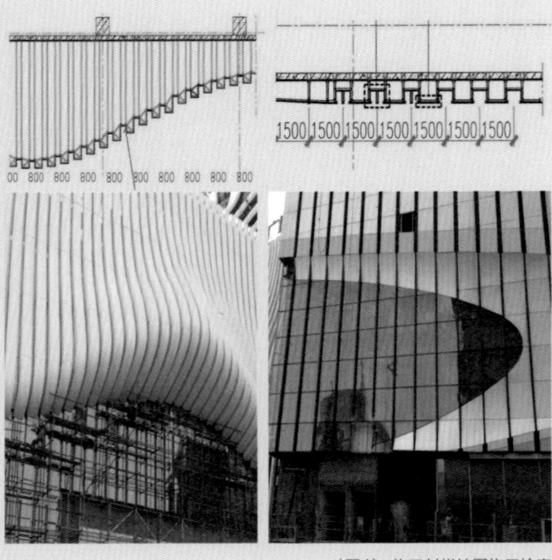

（图4） 施工封样按图施工检查

（图5） 现场实施效果

三、分工组织协调

现场管控的实施主体分为万达总部、项目公司、设计供方、总包单位四类。每一管控单位又分为不同的专业和工种；同时，不同系统对同一管控内容存在交叉管理模式。因而，设计成果的高效转化与推进，离不开有效的分工、组织与协调工作。分清权责界面、有效分工，是理清多头组织工作的有效手段。

在材料封样阶段，项目公司并依据设计封样对材料封样进行审核并签确；规划院审核项目公司上报的材料封样，审核确认后移交总包单位；设计供方在初设及施工图阶段进行材料封样，材料封样是对设计封样的深化，要求与设计封样保持一致；总包单位接收项目公司移交的材料封样，并以材料封样为依据进行样板段阶段的施工封样的选择。

在施工样板段阶段，规划院对项目公司报备的施工封样进行备案，并抽查考核；项目公司组织设计单位对样板段施工封样进行审核，并对最终施工封样进行签确。项目公司对签确的施工封样报备规划院；设计供方配合项目公司对总包单位的施工封样进行审核，并签字确认；总包单位根据材料封样选择多种材料进行样板段施工封样。总包单位根据项目公司签确的施工封样进行施工。

四、现场管控成果备案及评价考核

现场管控成果均应形成有效文件，以流程形式或实体形式及时备案。集团根据上报材料，对设计的实际效果、材料封样及样板段的实施成果进行评价。其中成功的经验作为优秀案例向各部门推广，并作为现场管控标准修订的依据。

the effective transformation and advancement of design results can hardly be achieved without effective division of labor, organization and coordination. To draw a clear distinction between rights and liabilities and to effectively divide the work are effective measures for sorting out multifarious organization work.

In the stage of sample material sealing, the project company shall check and approve the sealed sample materials with signature based on the designed sealed sample materials; the Planning Institute shall further check the sealed sample materials submitted by the project company, and after check and approval, the Planning Institute shall hand them over to the main contractor; in initial design and construction drawing design stage, the design provider(s) shall further specify and complement the sealed sample materials, which shall be a detailed version of and in accord with the designed sealed sample material; the main contractor, upon the acceptance of the sealed sample materials handed over by the project company, shall carry out the selection of sealed sample used in the construction of mock-ups on the basis of the approved sealed sample materials.

In the stage of sample section construction, the Planning Institute shall put the construction sealed sample materials provided by the project company on records, and carry out spot check; the design institute(s) shall check the construction sealed sample materials used for sample section construction under the guidance of the project company, and approve the final construction sealed sample with signature. The project company shall inform the Planning Institute of the approved and signed construction sealed sample; the design provider(s) shall cooperate with the project company's attempts to check the construction sealed sample provided by the main contractor, and approve with it signature; the main contractor shall choose various materials as the sealed sample for sample section construction based on the approved sealed sample materials.

IV. FILING OF ONSITE CONTROL ACHIEVEMENTS AND ASSESSMENT

Effective documents shall be made concerning the onsite control achievements, and shall be put on records in time report protocol or in a substantial form. Based on the submitted documents, the Group will evaluate the actual effects of the design, the sealed sample materials and the

集团部门依据相关考核办法对项目公司、设计供方和总包的现场管控进行考核，及时纠正现场管控中存在的问题与不足。

在材料封样及样板段阶段，材料封样经项目公司审核签确流程上报规划院审核；审核确认后移交总包单位，实体样板须存档；施工封样须经设计供方、项目公司、规划院签字确认后报备规划院。规划院对材料封样、施工封样及样板段进行合规性考核。

万达集团及项目公司现场正式规划设计检查成果均应形成会议纪要及销项清单，经相关各方签字确认后，流程备案及实体存档。对安全问题需落入安全管理系统，并督办总包单位完成整改，然后落入总包履约评估。其中，设计类问题，纳入对设计单位履约评估；现场管理类问题，纳入对项目公司的考核。

在万达集团的持有物业全过程技术管控中，现场管控对于自持物业设计成果转化具有举足轻重的作用。万达规划在项目现场管控中脚踏实地、推陈出新，紧随着集团第四次转型的深入开展，以坚实的积淀和锐意进取的创新精神，在集团飞速发展的进程中跟上时代的步伐，成为万达集团新的核心竞争力。

implementation achievement of mock-ups construction. The successful experience thereof, considered as outstanding cases, shall be spread to all departments, and serve as the basis for the compilation of Onsite Control Standard.

Departments of the Group shall carry out evaluation of the onsite control performance of the project company, the design provider(s) and the main contractor in accordance with relevant evaluation methods, and correct the existing problems and deficiencies of onsite control without any delay.

In the stage of sample material sealing and mock-up construction, the sealed sample after being checked and approved with signature by the project company shall be submitted to the Planning Institute for check and approval; after being approved, the sample shall be handed over to the main contractor and substantial sample plate shall be retained in the archive; the sealed sample used for construction after being approved with signatures by the design provider(s), the project company and the Planning Institute shall be submitted to the Planning Institute. The Planning Institute shall carry out compliance check and assessment on the sealed sample material, the sealed sample for construction and the mock-ups.

Meeting summaries and item lists shall be made concerning the assessment results of the onsite official planning and design of the Wanda Group and the project company, and after being approved with signatures by relevant parties, they shall undergo the report protocol and be retained in the archive. The Safety Management System shall be responsible for safety issues and supervise the main contractor to carry out rectification, before being evaluated by the main contractor performance system. Wherein, issues concerning design shall be classified into the Design Institute Performance Evaluation System, while issues concerning onsite management shall be classified into the Project Company Performance Evaluation System.

万达商业规划

持有类物业　上册 VOL.1

WANDA COMMERCIAL PLANNING 2014
PROPERTIES FOR HOLDING

2014

朱其玮　吴绿野　王群华　叶宇峰　任志忠　兰勇　兰峻文　张涛　黄勇
赖建燕　孙培宇　张琳　曹亚星　吴晓璐　苗凯峰　徐小莉　尚海燕
李文娟　刘婷　安竞　马红　曹春　侯卫华　张振宇　范珑　杨旭
谷建芳　张振宇　李淑仪　叶甲刚　雷磊　王鑫　李彬　张鹤　张飚
毛晓虎　莫鑫　都晖　刘江　蓝毅　郝宁克　屈娜　冯腾飞　张宝鹏
邵汀潇　万志斌　孙佳宁　袁志浩　阎红伟　吴迪　李斌　徐立军
宫赫谣　王雪松　张立峰　陈维　谢冕　刘杰　党恩　高振江
孙海龙　沈余　李昕　李海龙　罗沁　周澄　孙辉　齐宗新　黄引达
刘冰　杨艳坤　潘立影　程欢　邓金坷　康斌　刘易昆　李浩然
李江涛　钟光辉　张宁　黄春林　黄国辉　张洋　石路也　孟祥宾
刘阳　刘佩　耿大治　章宇峰　陈杰　冯志红　谷强　李小强　葛宁
张鹏翔　虞朋　田中　李洪涛　吕鲲　康宇　王治天　朱岩　董根泉
任腾飞　王吉　沈文忠　张珈博　张震　刘洋　胡存珊　马逸均
李光　郭晨光　朱迪　王锋　谢杰　李志华　宋锦华　刘锋
方文奇　秦鹏华　杨东　张堃　凌峰　李涛　张宇　易帆　任洪生
李明泽　陈勇　刘刚　刘宵　郭雪峰　赵洪斌　孔新国　陈嘉　王玉龙
刘志业　李浩　冯董　黄路　曹彦斌　周德　张剑锋　肖敏　李易　段堃
闫颇　朱欢　唐杰　刘潇　张雪晖　熊厚　王静　黄川东　董明海　王凡
谢云　王昉　黄涛　锡锋　李捷　解放　庞博　关发扬　赵青扬
任意刚　张争　张志斌　辛欣　罗贤君　郭杨　傅博　李梦雷　赵陨
杨春龙　路滨　张顺　王少雷　汪家绍　顾梦炜　姜云娇　江智亮
白宝伟　王凤华　贺明　李健　卫立新　冯晓芳　庞庆　何志勇
宋永成　谭訸　郭宇飞　高杉楠　卜少乐　刘海洋　韩冰　高峰
王睿麟　王凯　王宝柱　野天星　王瑶　葛朗　张佳　王晓　徐春辉
王永磊　李常春　曹国峰　于崇　张龇　杨汉国　赵剑利　王文广
张永战　李晓山　罗冲　王权　张旭　赵晓萌　高达　方伟　刘俊
康冠军　陈海亮　晁志鹏　邹洪　郑鑫　周永会　陈志强　桑伟
张德志　陈涛　张宇楠　高霞　王清文　王俊君　吴凡　张黎明　谭瑶
张绍哲　汤英杰　全永强　钱昆　路清淇　刘安　林彦　康兴梁
陈晓州　白宇　白夜　周明　崔勇　陈理力　杨娜　杨华　韦云　马辉
刘昕　金柱　王朝忠　罗琼　洪斌　赵宁　刘晓波　韩博　张烁君
徐广揆　魏大强　金博　马骁　程波　王鹏　柏久绪　闵盛勇　朱广宇
蒲峰　杜晶晶　汤钧　主佳　张浩　李扬　孟昆廷　赵海滨　钟文渊
王云　王奕　梁超　李丽　张雁翔　余斌　陈玭潭　韩天　宋雷　张述杰
王进纯　马雪健　李达　李万勇　耿磊　王政　王翔　张啸鸿
马长宁　姚建刚　万勇　李韦达　杨琳　马刚　王连发　殷超　刘向阳
李典　曹羽　陶晓晨　李民伟　张晓冬　法尔科内·马利亚　李华
卡斯特罗·索菲娅　季文君　刘怡德　夏海青　诺贝·马科斯　马佳
李云　张伟　孙穆元　吴科帆　武振衡　赵良颖　罗强　弓永康
廖伟　赵明　冯科力　张德志　陆宇亮　卫婷　林涛　曹玲妹
马嘉岳　柴刚军　孟晗　栾海　陆峰　林彬　王宇石　赵旭千

（以入职先后为序）

周大福
Jeep

萬達商業規劃

图书在版编目（CIP）数据

万达商业规划 2014 : 持有类物业 / 万达商业规划研究院主编 .
北京 : 中国建筑工业出版社 , 2015.10
ISBN 978-7-112-18634-1

Ⅰ. ①万… Ⅱ. ①万… Ⅲ. ①商业区—城市规划—中国 Ⅳ. ① TU984.13

中国版本图书馆 CIP 数据核字 (2015) 第 256490 号

责任编辑：徐晓飞　张　明
执行编辑：康　宇
美术编辑：康　宇　马雪健　陈　唯
英文翻译：喻蓉霞　王晓卉　郝　婧
责任校对：刘梦然

万达商业规划 2014：持有类物业
万达商业规划研究院　主编
*
中国建筑工业出版社出版、发行（北京西郊百万庄）
各地新华书店、建筑书店经销
北京雅昌艺术印刷有限公司制版
北京雅昌艺术印刷有限公司印刷
*
开本：787×1092毫米　1/8　印张：58　字数：1500千字
2015年10月第一版　2015年10月第一次印刷
定价：1180.00元（上、下册）
ISBN 978-7-112-18634-1
（27941）